The event safety guide

A guide to health, safety and welfare at music and similar events

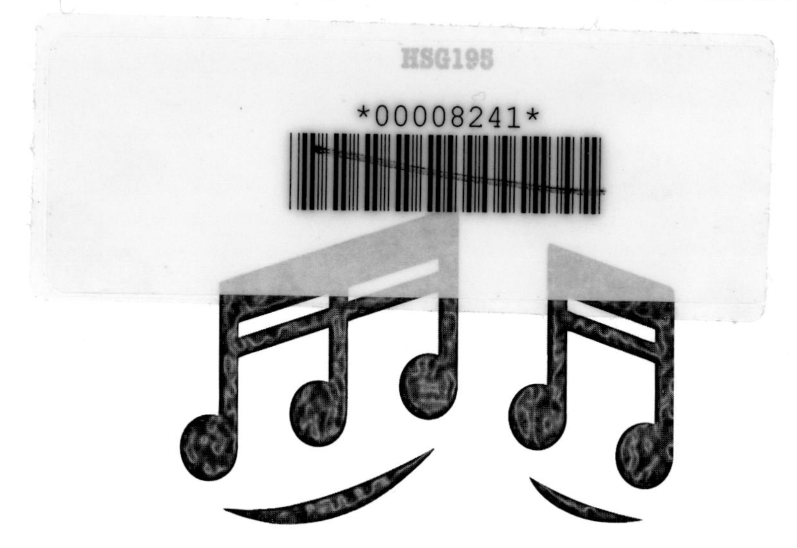

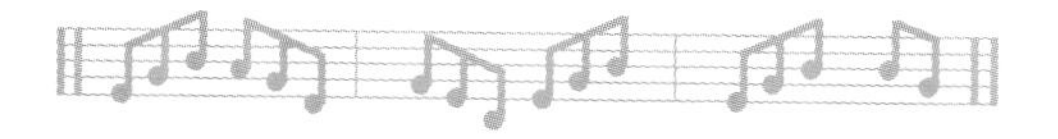

HSE BOOKS

First published as *Guide to health, safety and welfare at pop concerts and similar events* 1993
ISBN 0 11 341072 7

Second edition published 1999 ISBN 0 7176 2453 6
Reprinted 2000, 2001

This guidance is issued by the Health and Safety Executive. Following the guidance is not compulsory and you are free to take other action. But if you do follow the guidance you will normally be doing enough to comply with the law. Health and safety inspectors seek to secure compliance with the law and may refer to this guidance as illustrating good practice.

Contents

Foreword

In 1993, HSE, in conjunction with the Home Office and the Scottish Office, published *The guide to health, safety and welfare at pop concerts and other similar events*. The publication was well received and was adopted as the standard for planning and managing health and safety at these events.

In the light of changes to health and safety law, and the continuing development of best practice by the event industry, the guide has been reviewed and revised. The advice in this publication has been tried and tested and lays down a firm foundation on which to organise health and safety at music events. Many of the chapters can be applied to other types of event which, although not necessarily having a musical theme, share some of the common characteristics of music events.

This publication will enable event organisers, local authorities, the emergency services and HSE to work together to improve event safety. Their commitment to the production of this guide will ensure that health and safety remains a priority and that all involved will be able to continue to enjoy these events in safety.

This guide has been written in consultation with an event industry working group and with the co-operation of a great many people. All contributors, including the members of the working group, are listed at the back of the publication in the *Acknowledgements* section. I am grateful to them for the time, knowledge and expertise which was given freely and without which, this guide would not be possible.

I would like to give particular thanks to Lorraine Miller-Patel, Senior Environmental Health Officer, London Borough of Haringey, who had the unenviable task of co-ordinating the work of this project.

Jenny Bacon
Director General, 1999

Introduction

About this guide and its aims

1 The guide aims to help those who organise music events so that the events run safely. The event organiser, whether an individual, collective or local authority, has prime responsibility for protecting the health, safety and welfare of everyone working at, or attending, the event.

2 The guide brings together information needed by event organisers, their contractors and employees to help them satisfy the requirements of the Health and Safety at Work etc Act 1974 (HSW Act) and associated regulations. It will also enable organisers to understand the needs of others concerned with events, such as the local authority and emergency services, with whom they will need to co-operate.

3 The guide makes clear what is required and why it is necessary or sensible to do this. The guide does not replace the need for event organisers to seek advice from other sources and particularly for consultation with local authorities and emergency services.

4 As well as event organisers, others will find the guide useful, eg local authorities, health and safety enforcement officers, emergency services, contractors and subcontractors working at the event. The guide provides basic standards and safety measures through which it is hoped to encourage a consistency of approach while leaving scope for flexibility, taking into account the nature and size of the event.

How this guide differs from the previous edition

5 The guide is based on the previous edition, but has been updated to reflect changes in legislation, technology and working methods. Following consultation, it has been broadened to cover a wider range of topics and types of events. Some of the new topics include merchandising, camping and amusements.

6 This guide focuses on the application of the HSW Act and associated regulations and not the application of public entertainment legislation. Entertainment licensing authorities may, however, refer to this guide when considering appropriate entertainment licence conditions.

Applying the guide to event types

7 The guide is based on the principles of health and safety management and risk assessment. These acknowledge that each event will be different and will require a particular configuration of elements, management, services and provisions.

8 The guidance offers useful suggestions for many types of music event that take place at a variety of venues such as purpose-built arenas, sites not designed for public entertainment, open-air stadia, parks and greenfield sites. It is not, however, primarily intended to be applied to nightclubs and discotheques.

9 The type of music event may vary enormously, including rock, classical, traditional, contemporary and world music. Events vary in size and complexity from a concert of 500 local residents, to a major festival lasting several days and attracting in excess of 100 000 international visitors. All are covered by legal requirements, but the arrangements that may be needed will vary according to the event.

How the guide is arranged

10 Good planning and management are fundamental to the success of any music event. The first chapter of the guide gives event organisers essential points to consider in these areas as well as general advice on legal duties.

11 Subsequent chapters provide advice on specific arrangements for the health and safety of those involved in events, including the provision of services and facilities. There are also chapters which give some specific guidance for different types of event. These chapters should not, however, be read in isolation of all other chapters. The final chapter outlines issues relating to employees and other workers and provides a summary of the law relating to events.

12 Where other guidance is available, event organisers are recommended to refer to this. Technical details contained in the Home Office's *Guide to fire precautions in existing places of entertainment and like premises* and the Institution of Structural Engineers document, *Temporary demountable structures: Guidance on design, procurement and use* are not repeated in this guide.

13 All event organisers are recommended to use the chapter headings as a checklist for planning the requirements for their event. By applying a risk assessment approach to the type and size of event, it should be straightforward to decide which elements from each chapter are relevant and to assess the level and type of provisions needed at a particular event.

Chapter 1

Planning and management

14 In order to protect the health, safety and welfare of people attending a music event, as well as the employees, contractors and subcontractors working at the event, health and safety has to be managed. It is of fundamental importance to appreciate that planning for effective health and safety management should start at the same time as the planning for all other aspects of the proposed event.

15 The event organiser for the purposes of this publication is the individual or organisation who promotes and manages an event. More detailed information concerning the responsibilities of event organisers can be found in the chapter *Health and safety responsibilities*.

16 The aim of this chapter is to help event organisers plan for and manage their event safely. It explains the principles that underpin good health and safety management and sets out a basic approach that event organisers may wish to adopt to manage safety at events.

Health and safety management

17 The key elements of successful health and safety management include:

- creating a health and safety policy;
- planning to ensure the policy is put into practice;
- organising an effective management structure and arrangements for delivery of the policy;
- monitoring health and safety performance;
- auditing and reviewing performance.

Health and safety policy

18 A safety policy is a document that demonstrates to others that the company or organisation to which it relates accepts that concern for health and safety is an integral part of its organisation at all levels and that the highest management within the company mean to ensure that this concern will be translated into effective action. In other words, it is a way of letting others know your commitment to health and safety. This information is conveyed in the policy statement.

19 Safety policies should also contain details of the *organisation*, which show how the policy will be put into practice. This part will describe the roles and responsibilities of other people that have been given safety duties (not ultimate responsibility as this cannot be delegated). The organisation section of the safety policy should contain other matters, eg a diagram showing the delegation of safety duties, the nomination of people with the authority and competence to monitor safety and the resources (in time and money) that are available for health and safety.

20 The *arrangements* cover the detailed matters, eg the maintenance of a safe place of work, safe systems of work, safe access, provision of information, training and consultation with employees.

21 It is a legal requirement for employers employing five or more people to produce a written health and safety policy. Further information on safety policies can be found in HSE's booklet *Writing a safety policy statement: Advice for employers.*

22 The event organiser may be a person or organisation that promotes and manages an event themselves, eg promoters, production companies or local authorities. If you fall into this category, it is likely that you will have more than five employees and are legally required to produce a safety policy for the event. If you have been hired to promote and manage an event on behalf of another company or organisation, eg a client, you may not actually be an employer or have any employees. However, it will still be necessary to establish who has the overall responsibility for complying with the Health and Safety at Work etc Act 1974 (HSW Act) and to ensure that the responsibilities are recorded.

23 Some music events may be organised by people or organisations where there is no actual employer, eg community events, so there will be no legal requirement to produce a safety policy. However, there is still the legal responsibility for the management of contractors and subcontractors on site. Producing a safety policy in these circumstances is recommended as it provides a framework around which you can manage health and safety at the event.

24 The health and safety policy could relate to a series of events if these are to be organised by the same event organiser. An event health and safety policy prepared for a series of events will need to be reviewed in terms of the *organisation* and *arrangements* for health and safety for each particular event.

25 It is important that the safety policy details a management structure which defines the hierarchy of health and safety responsibility for the duration of the event and that these details are recorded in the safety policy document. (The duration of the event starts at the beginning of the build-up through to the finish of the breakdown.)

26 If an event is to be staged in existing premises such as an arena or a sports stadium, the event organiser will need to liaise with the venue or ground management in relation to the existing arrangements for health and safety.

Planning for safety

27 Effective planning is concerned with prevention through identifying, eliminating and controlling hazards and risks. The amount of time that needs to be set aside for planning will be very much

dependent upon the size, type and duration of the music event. For large events, experience shows that 6-9 months beforehand is not too early to start.

28 Other chapters in this publication give specific advice and guidance in their subject area. It is therefore necessary to have an appreciation of the information contained in all chapters to be able to plan effectively.

The phases of an event

29 The planning issues for an event can be considered in separate parts:

- the 'build-up', which involves planning the venue design, selection of competent workers, selection of contractors and subcontractors, construction of the stages, marquees, fencing, etc;
- the 'load in', which involves planning for the safe delivery and installation of equipment and services which will be used at the event, eg stage equipment used by the performers, lighting, public address (PA) systems, etc;
- the 'show', which involves planning effective crowd management strategies, transport management strategies and welfare arrangements. Planning strategies for dealing with fire, first aid, contingencies and major incidents are important ;
- the 'load out', requires planning for the safe removal of equipment and services;
- the 'breakdown', which includes planning to control risks once the event is over and the infrastructure being dismantled. Collection of rubbish and waste-water disposal present risks and these aspects need to be planned and managed.

Planning for the build-up

30 To minimise risks during the build-up, ensure that the venue is designed for safety (see chapter on *Venue and site design*). It is also necessary to ensure that any infrastructure which will be used at the event, such as stages, seating, tents, marquees or other structures will be erected safely and be structurally safe once erected and used (see chapter on *Structures*).

31 Prepare plans to show the location of the stages, barriers, front-of-house towers, delay towers, entries and exit points, emergency routes, first-aid and triage areas, positioning of toilets, merchandising stalls, etc. It may be necessary to obtain plans of existing premises from the owner, occupier or venue manager in which your event is to be held. Copies of these plans may need to be given to the contractors building the infrastructure to ensure correct positioning of the various structures to be used at the event.

32 Ask contractors and subcontractors to provide copies of their own health and safety policies, and details of any hazards and risks associated with their work, before the build-up commences. Documents and calculations will also need to be obtained in relation to the stages, seating or other temporary demountable structures. These plans, documents, and calculations will be needed when discussing your event with health and safety inspectors, local authority licensing officers and officers of the emergency services.

33 Plan the arrival of the contractors and ensure that their activities on site are co-ordinated with others. Also plan the provision of first aid and welfare facilities for the people who will be working on site, and ensure that they are suitable, in sufficient numbers and available from the time that work begins.

34 It is good practice to draw up a set of site safety rules and communicate these rules to the contractors before or as soon as they arrive on site. They can be posted in the form of signs in site offices and other areas. Contractors will then be aware of safe working practices required of them at the particular site or venue.

Planning for the load-in

35 Once the infrastructure has been built all other equipment and services will need to be brought to the site and installed in or on the structures, eg the loading of the performers' equipment onto the stage (which is likely to involve manual handling procedures) and the delivery of equipment to be used in the bar areas. These operations will also need careful planning.

Planning for the show

36 Planning for the show requires preparing strategies for crowd management, transport management, fire, first aid, major incident and contingency planning. More specific details about planning these aspects can be found in other chapters later in this publication. Successful planning for the show requires a team approach. It cannot be achieved by one individual operating alone but requires seeking information and advice from the emergency services (such as the police, fire brigade, etc), the health authority, local authority, any existing venue managers, stewarding, and security contractors.

37 Create an event safety management team to co-ordinate the planning aspects of the show itself. The event safety management team could include members of the local authority and emergency services. It may also be advisable to set up a series of safety planning meetings so that information can be exchanged between the parties and to ensure that the relevant agencies are aware of the planning process. Table-top emergency planning exercises to test the validity of the emergency plans for the larger and more complex events may also be useful.

The event safety management plan and event safety team meetings

38 To provide a comprehensive overview to all these planning aspects it may be helpful to produce an event safety management plan. The constituents of an event safety management plan could include the following:

- the event **safety policy** statement detailing the organisation chart and levels of safety responsibility;
- the event **risk assessment** (see paragraphs 41-48);
- details of the event including venue design, structures, audience profile and capacity, duration, food, toilets, refuse, water, fire precautions, first aid, special effects, access and exits, music levels, etc;
- the **site safety plan** detailing the site safety rules, site crew managers and safety co-ordinator, structural safety calculations and drawings;
- the **crowd management plan** detailing the numbers and types of stewards, methods of working, chains of command;
- the **transport management plan** detailing the parking arrangements, highway management issues and public transport arrangements;
- the **emergency plan** detailing action to be taken by designated people in the event of a major incident or contingency;
- the **first-aid plan** detailing procedures for administering first aid on site and arrangements with local hospitals.

39 Remember that the constituents of the event safety management plan are your working documents and will need to be reviewed and updated as new information is received either before or during the event. It is only necessary to produce this plan for the key members of your event safety team. Ensure that there is full document control so that redundant or superseded documents are not mistaken for the final version.

40 Event safety planning meetings are an ideal way to ensure that the event safety management team members are updated on the content of the plan, as well as providing a mechanism for ensuring a flow of safety information on a regular basis. These meetings can be arranged in the weeks or days leading up to the event. If the event is to take place over a few days, eg festivals, meetings should take place at least once each day of the event.

The event risk assessment

41 The Management of Health and Safety at Work Regulations 1992 (Management Regulations) require all employers and self-employed people to assess the risks to workers and others who may be affected by their work.

42 The purpose of a risk assessment is to identify hazards which could cause harm, assess the risks which may arise from those hazards and decide on suitable measures to eliminate, or control, the risks. Significant findings of the risk assessment must be recorded if five or more people are employed. A risk assessment for the build-up, show and breakdown, can only be carried out once information has been received from the contractors, other companies and self-employed people who will be working on site. It will also be necessary to visit the site or venue to identify specific hazards.

43 A *hazard* is anything which has the potential to cause harm to people. This could be a dangerous property of an item or a substance, a condition, a situation or an activity.

44 *Risk* is the likelihood that the harm from a hazard is realised and the extent of it. In a risk assessment, risk should reflect both the likelihood that harm will occur and its severity.

45 Hazards associated with the assembly of large numbers of people may vary according to the nature of the event and these hazards should be similarly assessed in terms of risk. The previous history of the performers and the audience that they attract can provide valuable information. The overall event risk assessment will then indicate areas where risks need to be reduced to acceptable levels.

46 There are five steps which need to be taken to assess the risk associated with staging the event.

Step 1	Identify the hazards associated with activities contributing to the event, where the activities are carried out and how the activities are to be undertaken
Step 2	Identify those people who may be harmed and how
Step 3	Identify existing precautions, eg venue design, operational procedures or existing 'safe systems of work'
Step 4	Evaluate the risks
Step 5	Decide what further actions may be required, eg improvement in venue design, safe systems of work, etc

47 The risk assessment findings will need to be recorded and a system developed to ensure that the risk assessment is reviewed and, if necessary, revised.

48 Further helpful information on how to carry out a risk assessment can be found in the documents *Research to develop a methodology for the assessment of risks to crowd safety in public venues* and the publication *Five steps to risk assessment.*

Planning for the load out

49 Although the music event has ended, this does not mean that the responsibilities towards health and safety are over. Ensure that you have considered how the equipment and services will be removed from the stages, tents and marquees at the end of the event.

Planning for the breakdown

50 The stages, marquees and stalls have to be dismantled safely and in a controlled manner and removed from site. Plan to ensure the same site safety rules apply in relation to managing contractors during this phase of the event.

Organising for safety

51 Once the health and safety policy statement has been prepared and the levels of responsibility have been agreed and you have prepared your safety plans, it is necessary to organise for safety especially when work is to begin on site.

52 Effective organising contains these four elements:

- competence
- control
- co-operation
- communication.

Competence

53 Competence is about ensuring that all employees, self-employed people, contractors and subcontractors working on your site have the necessary training, experience, expertise and other qualities to carry out the work safely. Competence is also about ensuring the right level of expertise is available, particularly in relation to specialist advice.

54 Ensure that the contractors or subcontractors you intend to hire, to build the infrastructure or provide other services, are competent in the management of their own health and safety when working on site. Simple checks of the contractors' and subcontractors' health and safety policies can be carried out and applicable safety method statements and risk assessments obtained and examined in relation to their proposed work.

Control

55 Establishing and maintaining control is central to all management functions. Control starts with the production of a health and safety organisational structure, which details specific health and safety responsibilities and shows clear reporting mechanisms. Control also ensures that the contractors and self-employed people understand their responsibilities and that they know what they must do and how they will be held accountable for safety on site. It is important to make sure that contractors understand how health and safety will be controlled and monitored before they begin work on site.

Co-operation

56 Effective co-operation relies on the involvement of employees, contractors, and others, in your planning, standard setting, operating procedures and instructions for risk control as well as involvement in monitoring and auditing. Co-operation enables the risks to be suitably controlled by allowing the exchange of information.

57 Contractors, subcontractors and self-employed people need to appreciate the hazards and risk to others working on site and to co-operate with each other to minimise identified risks. Effective co-operation can be achieved by working to prepared site safety rules and safety plans.

Communication

58 Effective communication ensures that all those who are to work on site understand the importance and significance of the health and safety objectives. Make sure that you keep contractors, subcontractors, and others informed of safety matters and procedures to be followed on site.

59 Further helpful information about managing contractors on site can be found in the HSE publication *Managing contractors: A guide for employers*.

Monitoring safety performance

60 Monitoring is essential to maintain and improve health and safety performance. There are two ways of generating information on safety performance:

- active monitoring systems; and
- reactive monitoring systems.

61 Active monitoring systems give feedback on safety performance before an accident or incident happens. Active monitoring can be achieved by carrying out inspections of the contractors on site during the build-up and breakdown and by checking the contractors' safety method statements for carrying out work against their actual work on site.

62 Reactive monitoring systems are triggered after an accident or incident has occurred. They include identifying and reporting injuries, ill health, other losses such as damage to property, incidents with the potential to cause injury, and weaknesses or omissions in safety standards.

63 Information obtained during inspections as well as a result of incidents or property damage can be recorded in an event logbook. This book can be used to keep other records and the information used to audit and review the event at a later date. Without information from both systems, it would be impossible to assess your safety performance against your safety standards set in the safety policy. It therefore follows that without these monitoring systems no improvements in safety performance would take place for future events.

The role of the safety co-ordinator

64 Event organisers must have access to competent help in applying the provisions of health and safety law unless they are competent to devise and apply protective measures themselves. A competent person is someone who has sufficient training, expertise, experience or knowledge and other qualities that enable that person to devise and apply protective measures.

65 Appoint a suitably competent safety co-ordinator to help you comply with health and safety legislation and ensure that the safety co-ordinator reports directly to you. Safety co-ordinators can assist in the:

- selection and monitoring of contractors;
- liaison with contractors, self-employed people on site and the health and safety enforcement authority;
- checking of safety method statements and risk assessments;
- preparation and monitoring of site safety rules;
- checking of appropriate certificates in respect of structures, electrical supplies, etc;
- communication of safety information to contractors on site;
- monitoring and co-ordinating safety performance;
- co-ordinating safety in response to a major incident.

66 To be effective, the safety co-ordinator needs to have access to the safety documentation supplied by the contractors. The safety co-ordinator also needs to be easily available to workers on site from the beginning of the build-up of the event through to the final breakdown. The safety co-ordinator should also be a member of your event safety management team.

67 It is not recommended that event organisers appoint themselves as the safety co-ordinator. To be effective the safety co-ordinator should not have other competing roles which would inevitably face an event organiser during the course of the event.

Auditing and reviewing safety performance

68 Auditing aims to establish that appropriate safety management arrangements are in place, adequate risk control systems exist and that they are being put into practice. Carry out auditing at the completion of every music event so that any problems identified in your planning, organisation or any matters that arise during the event can be analysed and corrected for any future events. Views of the police, fire brigade, health authorities, first-aid providers and local authority can be sought as well as views of the safety co-ordinator, contractors and stewarding contractors.

69 Arrange for a debriefing after the event to review the effectiveness of the safety management systems. The local authority may also ask you to attend a debriefing meeting so that they can give you some feedback on your safety management systems from their perspective.

Liaison with the local authority and emergency services

70 The local authority will usually request a preliminary meeting so that the proposals for the event can be discussed. Members of the emergency services as well as health and safety inspectors may attend. It may be helpful to ask the local authority to provide you with a checklist of information required for prior approval along with the timescale for submitting the information. The information you supply should be sufficient to enable the local authority to examine your safety management systems and check any necessary plans, calculations and drawings.

71 Local authorities will not usually require a copy of every safety-related document in advance of the event unless considered necessary. They may, however, require evidence that you have planned your event safely before the event takes place. Ensure that any safety documentation is easily available for examination by health and safety inspectors or other local authority officers. Keep your information in a safety file as this would make this process easier and ensure that safety information is not misplaced. Make suitable arrangements so that the local authority can contact you quickly for matters that may need further clarification. Last minute changes are not conducive to good safety planning and management.

72 Continue to liaise with the local authority and members of the emergency services once the necessary permission to stage the event has been granted. Consider inviting these organisations to your event safety team meetings to ensure that they are updated on aspects of the event safety management plan.

The public entertainment licence and the HSW Act

73 It is usually necessary to obtain a public entertainment licence from the local authority for most music events. Permanent venues usually have an annual entertainment licence granted with specific conditions attached for different types of events. If you are organising an event in premises with an existing entertainment licence you will need to familiarise yourself with its specific requirements.

74 Public entertainment licences do not replace the need for you to comply with the provisions of the HSW Act. The aim of this publication is to help you comply with the provisions of the HSW Act and should not be confused with entertainment licensing which is dealt with under separate legislation. Information on entertainment licensing legislation can be found in the chapter on *Health and safety responsibilities*. Local authorities may, however, refer to this publication when considering appropriate entertainment licence conditions.

Chapter 2

Venue and site design

75 This chapter gives an overview of all the factors that need to be considered when designing your venue or site. More detailed information on certain topics can be found in other chapters of this publication.

76 The general principle behind venue design is to provide an arena in which the audience can enjoy the entertainment in a safe and comfortable atmosphere. The requirement for certain safety provisions, the type, number and specification of facilities and services will depend on the type of event and the outcome of the risk assessment (see paragraph 41-48).

77 The final design of a site will be dependent on the nature of the entertainment, location, size and duration of the event. It will also need to take account of the existing geographical, topographical and environmental infrastructure.

Site suitability assessment

78 It is important to visit the venue or site to carry out a preliminary assessment to determine suitability. The main areas for consideration are: available space for audience, temporary structures, backstage facilities, parking, camping and rendezvous points. You may already have a proposed capacity in mind, together with some ideas of the concept of the entertainment. Rough calculations of the available space are useful at this stage.

79 Factors to consider include the following.

- Ground conditions - are they suitable? Even and well-drained open sites are preferable. Avoid steep slopes and boggy areas.
- Traffic and pedestrian routes and emergency access and exits - what routes already exist? Are they suitable to handle the proposed capacity? Is a separate emergency access possible? If not, can other routes be provided? Are roads, bridges, etc, structurally sound? For further information see the chapter on *Transport management*.
- Position and proximity of noise-sensitive buildings - are there any nearby? Is it possible to satisfy both the requirements of the audience and the neighbours? A noise propagation test may be advisable.
- Geographical location - where is the site located? How far away is the hospital, fire station, public transport, parking, major roads, local services and facilities, etc? Such information can be valuable when assessing the suitability of the site and determining the extra facilities that need to be accommodated within the site.
- Topography - How does the land lie in relation to its surroundings? Does it form a natural amphitheatre? Where does the sun rise and set? Could any natural features assist in noise reduction? Are there any natural hazards/features such as lakes and rivers?
- Location and availability of services - water, sewage, gas, electric, telephone (including overhead cables). Are there any restrictions or hazards? Can they be used? Is the event site within the 'consultative distance' of a hazardous installation or pipeline?

80 The above aspects can be assessed by walking the site, studying the appropriate mapping and seeking advice and information from the land owner, local authority or venue management. Such information is essential before beginning detailed site design. For existing venues much of this information may be available from the venue management and/or local authority.

Pre-design data collection and appraisal

81 The next step in site design is to collect all the available data together and appraise it. The site design should be based on the site suitability and risk assessments.

82 Ensure that you have considered the following factors:

- proposed occupant capacity;
- artist profile;
- audience profile;
- duration and timing of event;
- venue evaluation;
- whether alcohol is on sale;
- whether the audience is standing, seated or a mixture of both;
- the movement of the audience between the entertainment and/or facilities;
- artistic nature of the event, single stage, multiple-arena complex, etc.

83 The above information can then be used to determine the provisions and facilities needed within the site, for example stages, tents, barriers, toilets, first aid, concessions, exits, entrances, hospitality area, sight lines, power, water, sewerage, gas, delay towers, perimeter fencing, backstage requirements, viewing platforms and waste disposal requirements. Once all the information is collated, detailed site design can begin.

Site plans

84 Once the basic outline has been determined, detailed scaled site plans should be produced. Often, many versions may be produced as amendments are made and as further information is obtained. Ensure, however, that your site plans are kept up to date and are given to members of your event safety team.

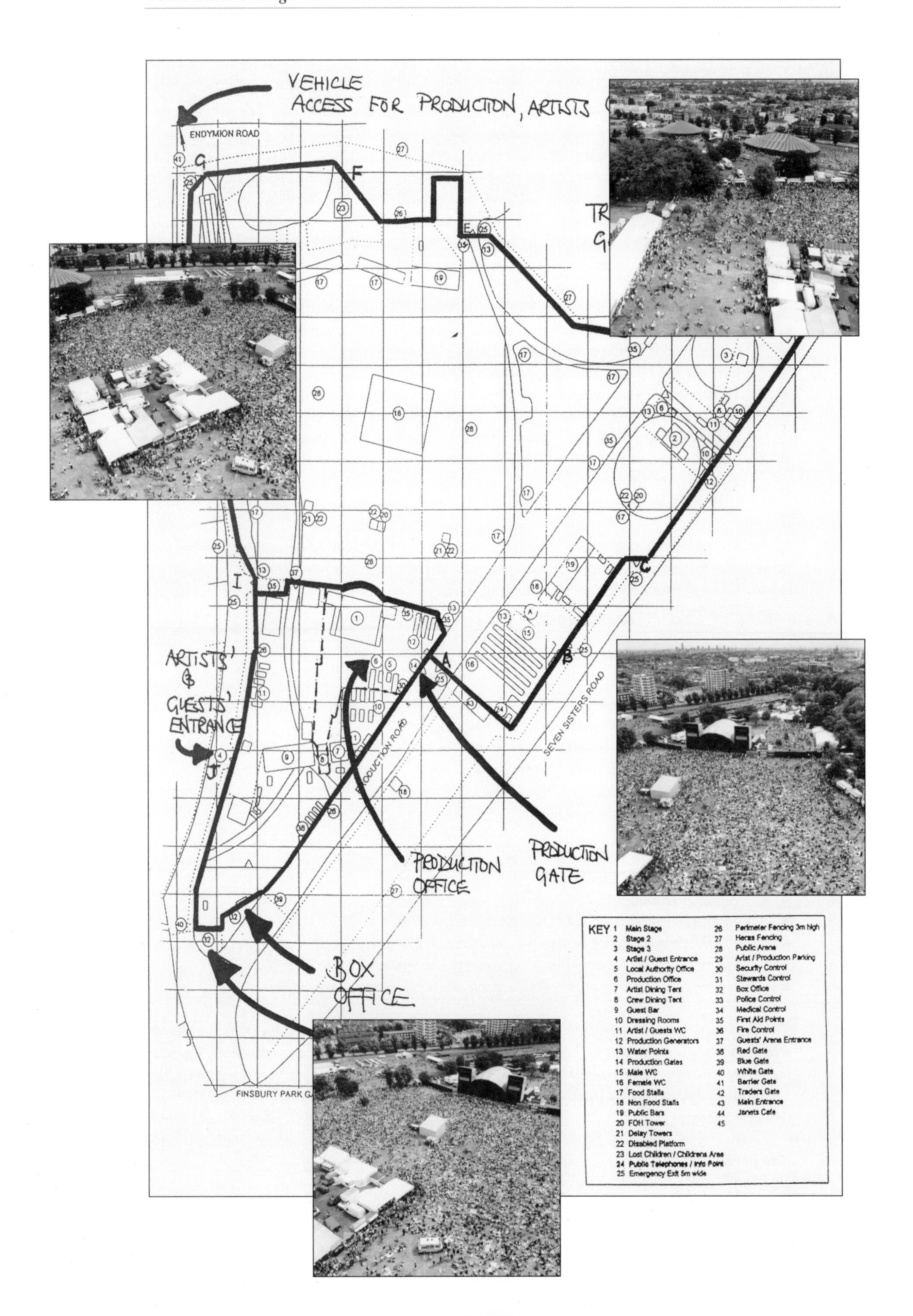
VEHICLE
ACCESS FOR PRODUCTION, ARTISTS
ENDYMION ROAD
ARTISTS' & GUESTS' ENTRANCE
PRODUCTION ROAD
SEVEN SISTERS ROAD
PRODUCTION OFFICE
PRODUCTION GATE
BOX OFFICE
FINSBURY PARK
KEY
1 Main Stage
2 Stage 2
3 Stage 3
4 Artist / Guest Entrance
5 Local Authority Office
6 Production Office
7 Artist Dining Tent
8 Crew Dining Tent
9 Guest Bar
10 Dressing Rooms
11 Artist / Guests WC
12 Production Generators
13 Water Points
14 Production Gates
15 Male WC
16 Female WC
17 Food Stalls
18 Non Food Stalls
19 Public Bars
20 FOH Tower
21 Delay Towers
22 Disabled Platform
23 Lost Children / Childrens Area
24 Public Telephones / Info Point
25 Emergency Exit 5m wide
26 Perimeter Fencing 3m high
27 Heras Fencing
28 Public Arena
29 Artist / Production Parking
30 Security Control
31 Stewards Control
32 Box Office
33 Police Control
34 Medical Control
35 First Aid Points
36 Fire Control
37 Guests' Arena Entrance
38 Red Gate
39 Blue Gate
40 White Gate
41 Barrier Gate
42 Traders Gate
43 Main Entrance
44 Janets Cafe
45

Make sure, however, that alterations are not made to the site plans after capacity levels have been determined and tickets placed on sale as the alterations may have an effect on sight lines and therefore available viewing areas. Plans may already exist for permanent existing premises.

Site-design considerations

Venue capacity/occupant capacity

85 The capacity of a venue is generally dependent upon the available space for people and the number of emergency exits. The latter is the subject of a calculation involving the appropriate evacuation rate, ie width of available exit space and appropriate evacuation route.

86 Some of the site will be taken up by structures which will be unoccupied. The rest of the site will need to be considered in calculating occupant capacity even though a direct view of the entertainment may not be possible for all locations. If there are any areas where the audience does not have a reasonable view of the performance, this space should be deducted from the available area or a lesser density used in calculations. Areas which could afford partial or total cover to the audience in the event of inclement weather, should be identified and the effects of audience migration to these areas considered.

87 In venues where seating is provided, the major part of the occupant capacity will be the lesser of the two figures determined by the number of seats and exit provision However, in other cases a calculation based on the acceptable occupant density should be carried out. Generally, 0.5 m^2 of available floor space per person is used for outdoor music events.

88 Double-check the preliminary occupant capacity calculation and exit requirements once all initial infrastructure requirements and facilities are in place on the site design. Further detailed information on occupant capacities can be found in the chapter on *Fire safety* and in the publication *Guide to fire precautions in existing places of entertainment and like premises.*

Exit requirements

89 The exit numbers for a venue depend directly on the occupant capacity and the appropriate evacuation time for the type of structure. The publications *Guide to safety at sports grounds* and *Guide to fire precautions in existing places of entertainment and like premises* give details which relate to stadia and fixed entertainment premises - they also provide sample calculations.

90 Place exits around the perimeter and ensure that they are clearly visible, directly and indirectly by signage. Ensure they are

free from obstruction on either side. The final exit terminus should be assessed and be as safe as possible, ie into open spaces, assembly areas, etc, rather than into a main road or into traffic flows. It is therefore important to examine these areas when carrying out your overall event risk assessment. Exit gates should operate efficiently and effectively. Where practicable provide separate exits for pedestrians, service and concession vehicles. Wheelchair access and exit will also need to be taken into consideration.

Venue access

91 Venue access is a function of the design and location of transport and parking facilities and the design of access roads. Such facilities have to be able to cope with the peak demand as determined from the arrival profile (see chapter on *Transport management*).

92 The layout of the access routes will obviously depend upon the location of facilities. Distribute routes around the site to minimise the load and ensure that the routes do not converge. The routes should be simple, easy to follow, direct and avoid cross flows.

Entrances

93 The entrances provide the means for supervising, marshalling and directing the audience to the event. At some venues they may be used as an exit, at others such as football stadia, they are separate. It may be necessary to provide separate entrances for performers, workers, guests, etc.

94 The design and location of entrances depends on the numbers of entrances required, where they are placed and the capacity to be handled at each entrance. There should be sufficient numbers of entrances to cope with the peak demand and achieve a smooth and orderly flow of people through them. The direction from which people are likely to come, the maximum number of people from each direction and the flow rate through the entrance are important issues which determine the number of entrances required. For purpose-built venues, these will already have been considered and approved.

95 Flow rates depend on the type, design and width of the entrances and whether or not searching takes place. (The *Guide to safety at sports grounds* gives flow rates applicable to sports stadia.) The desired entry time is the time taken to allow everyone access to the venue. This will depend entirely upon the type and duration of the event and the audience profile. The possibility of inclement weather may affect the desired time. Any queuing system to manage people at the entrance also needs to be planned and carefully designed.

Sight lines

96 It is important that the audience has a clear line of vision to the stage to avoid movement towards the centre. The widest possible sight lines help to reduce audience density in front of the stage and help to minimise surging and the possibility of crushing injuries. The stage width, height and position of PA wings, etc, all affect sight lines. Design sight lines to create areas of clearer space on the immediate stage left and right. This allows movement and emergency access.

Video screens

97 For very large audiences the increased distances between the stage and the back of the viewing area results in poor visibility and reduced entertainment value. This can lead to crushing and overcrowding. Strategically placed video or projection screens can be very effective. Screens located at some distance from the stage encourage a proportion of the audience to use a less crowded part of the site. Screens near the stage can help to stop people pushing towards it. Screens may require substantial foundations and support so sufficient space should be allowed in any site design. Not all types of screen operate in daylight and if the intention is to use a screen in these conditions, make sure that an appropriate type is used.

Seating arrangements

98 Where there is a risk of over-excitement among audience members, consider holding an all-seated event as this may help to prevent crowd surges and crushing at the front of the stage area. Spacing requirements and aisle widths, etc, can be found in the publication *Guide to fire precautions in existing places of entertainment and like premises*. If temporary seating is provided, seating will need to be adequately secured to avoid 'snaking'. Temporary seating must be approved by the local authority.

Slopes

99 Ensure that you have fully considered the effects of any slopes at your venue in your risk assessment. It may be necessary to consider providing exit steps or ramps with non-slip surfaces. The area in front of the stage should be as flat as possible to prevent tripping and crushing.

Observation points

100 At some outdoor music events, observation points may be considered necessary. These should be strategically placed to maximise the 'view' of the audience. Establish safe entrances and exits to these observation points.

Production infrastructure and backstage requirements

101 The production infrastructure will depend on the type, size and duration of the event. Typically, production offices, refreshment facilities, accommodation (for workers and artists), dressing rooms, storage space, equipment, etc, needs to be accommodated, usually backstage. Carefully consider the number of units required, fire hazards, access routes and circulation space, generators, first-aid posts, ambulance, fire and police requirements. Try to keep performers' areas separate from production and working areas.

Fire and ambulance requirements

102 Fire and ambulance requirements such as parking areas, first-aid posts, rendezvous points, triage areas, etc, need to be carefully assessed and positioned in the appropriate places. Design the site so that they are readily accessible and can be easily identified. Fire appliances should be able to access all parts of the site and be able to get within 50 m of any structure. Establish emergency access routes which are kept clear at all times. Temporary trackways may be necessary for wet, difficult ground. Consider separate gated entrances and exits, of sufficient height and width, for fire and ambulance vehicles.

Police and stewarding positions

103 The presence of police and the number and positioning of stewards will depend upon the nature and type of entertainment provided (see chapter on *Crowd management*).

Site workers

104 For large events a significant number of workers will be on site and will need their own facilities such as catering, toilets, showers, offices, sleeping accommodation, etc. Such facilities may form a separate compound or be distributed between backstage and/or main area. Carefully plan such requirements to incorporate them safely into the site design.

Hospitality area

105 The level of hospitality will vary with the size of the event. Accommodation and facilities may need to be provided for only a few people requiring no more than a small meeting area through to very large sophisticated complexes catering for several thousand people. Marquees and viewing platforms may be required. The exact requirements need to be planned and incorporated into the overall site and venue design. Often such large numbers are forgotten in the capacity calculations but need to be included.

Noise considerations

106 The overall site design and layout should maximise the audience's enjoyment and protect the neighbours from noise nuisance (see chapter on *Sound: noise and vibration*). Consider the stage location and other sound sources, in relation to nearby noise-sensitive properties and the topography of the site. Use slopes and natural barriers to their maximum effect. It may well be

advantageous to use a distributed sound system suspended from delay towers. Carefully consider the siting and construction of such towers to control sight lines, avoid crushing points and prevent unauthorised 'viewing' platforms.

Catering and merchandising

107 Position merchandising and catering concessions away from access routes and in less densely occupied areas of the arena. Some units will have highly flammable products such as LPG and require careful positioning (see chapters on food and drink and merchandising). Consider circulation space and potential queuing arrangements, which should not obstruct pathways.

Perimeter fencing

108 Whether or not a perimeter fence is required depends on the type and nature of the event. Fences may be necessary to prevent trespassers entering the site and for the safe management of the audience.

109 Some events may not require a fence, just a stake and tape/steward barrier, whereas others may need a sophisticated, substantial fence or multiple arrangement. Assess the crowd loading on such structures and the climbing potential.

110 A typical arrangement for large music events is an opaque inner fence with an outer fence - providing a moat in which stewards can patrol. To minimise the climbing of the inner fence for those who have breached the outer, a 5 m gap is usual to prevent the run-up approach. Three fences may be used which can easily form an emergency vehicle route. Carefully consider the ground conditions, obstructions, support legs and exit and entrance requirements.

Front-of-stage barrier requirements and arrangements

111 A front-of-stage barrier may be required particularly if significant audience pressure is expected. The risk assessment for the event, relating to the evaluation of the performer and audience profile, together with the capacity, should assist in determining whether or not one is required and if so what type and design is required. For most large music events, some form of front-of-stage barrier will be necessary (see chapter on *Barriers* for further information).

Signage

112 The location and size of all signage is critical when designing a site. For indoor/permanent venues such signage is normally in place for emergency exits, extinguisher points, entrances, car parks,

emergency vehicle points, etc. For supplementary facilities and all outdoor sites, this will not be the case.

113 The effective use of signs provides a rapid way of conveying orientation, directions and emergency information. It therefore assists in audience flow. Signage should be clearly visible and easily understood. Signs should be lit in the dark.

114 From a site-design perspective, the size and position is very important. Large outdoor venues will require signage larger than usual so that it can be seen from a distance. Fixture points may have to be constructed, such as scaffold towers, etc. Safety signs must conform to the Health and Safety (Safety Signs and Signals) Regulations 1996.

Welfare facilities

115 The number and type of welfare and information facilities, sanitary accommodation, water supply, etc, will depend upon the type of event, but once numbers have been agreed these need to be considered in your venue or site design.

116 Distribute sanitary accommodation around the site in a manner which does not block sight lines and serves the greatest need, eg near bars and catering concessions. If non-main units are to be used, plan access for the emptying tanker. Ensure they are clearly visible and well signed and that queuing areas do not obstruct any gate, emergency route, etc. Water supply is normally situated next to sanitary accommodation. If tankers are used, consider the space requirement and ground drainage.

117 Information points vary from a notice board to a marquee. The size and location must be taken into consideration. The best positions are near the main entrance into the site, but not too close to any gate or emergency access route, since people using or waiting near the facility could cause an obstruction. Try to locate welfare and information points in less noisy parts of the site.

Excess visitors

118 Contingency arrangements should be made to cope with excess visitors to an event. Measures may necessitate the design of a holding and/or queuing area and related facilities, which need to be accommodated within the design.

Final site design

119 Once all the necessary details and requirements have been finalised each should be drawn to scale on a site plan in relation to spacing requirements, etc. The final plan should then be reassessed to check the occupant capacity (in relation to sight lines and circulation space) and emergency services, worker and audience entry and exit. Power generation and distribution positions can now be finalised.

Chapter 3

Fire safety

Means of escape

120 The aim of this chapter is to explain what is necessary to ensure suitable and sufficient means of escape in case of fire for all people present. Further details are given in the *Guide to fire precautions in existing places of entertainment and like premises*, and advice may also be obtained from the fire authority for the area. Music events are subject to the requirements of both the Fire Precautions Act 1971 and the Fire Precautions (Workplace) Regulations 1997. The Fire Precaution (Workplace) Regulations 1997 also apply to any tent or moveable structure. In Scotland the means of escape may be subject to control under other legislation and the local authority must be consulted.

121 Whether the venue is in a building or outdoors it is likely that some adaptation may be needed to accommodate a music event. This chapter covers the means of escape which may need to be provided for buildings, sports stadia and at outdoor venues to safely accommodate a music event.

Definitions

122 The following definitions are used:

Final exit is the termination of an escape route from a building or structure giving direct access to a place of safety such as a street, passageway, walkway or open space and positioned to ensure that people can disperse safely from the vicinity of the building or structure and the effects of fire.

Means of escape is the structural means whereby a safe route is provided for people to travel from any point in a building or structure to a place of safety without assistance.

Place of safety is a place in which a person is no longer in danger from fire.

The occupant capacity is the maximum number of people who can be safely accommodated at the venue. In the case of standing areas at longer events there is a need to take into account 'sitting down' space for the audience and freedom of movement for access to toilets and refreshment facilities. It is essential to agree the occupant capacity with the local authority and fire authority as early as possible as the means of escape arrangements are dependent on this figure.

In areas where seating is provided, the major part of the occupant capacity will be determined by the number of seats available. However, in other cases, a calculation will need to be made and this is based on each person occupying an area of 0.5 m^2. The maximum number of people who can be accommodated can therefore be calculated by dividing the total area available to the audience (in m^2) by 0.5.

Example: an outdoor site measuring 100 x 50 m with all areas available to the audience could accommodate a maximum of 10 000 people (ie 100 x 50 m = 5000 m^2 divided by 0.5 = 10 000).

However, the local or fire authority may decide that for certain events the occupant capacity will need to be reduced.

General principles for means of escape

123 People should be able to walk to safety along a clearly recognisable route by their own unaided efforts regardless of where a fire may break out at the venue. However, for some people with disabilities it will be difficult, if not impossible, to make their way to a place of safety without the assistance of others. Consider carefully the arrangements for these people.

124 When evacuation is necessary, people often try to leave the way they entered. If this is not possible (perhaps because of the position of the fire or smoke), they need to be able to turn away from the fire and find an alternative route to a place of safety. However, the audience may underestimate the risk or be reluctant to use exits they are unfamiliar with. It is essential to train stewards to recognise this fact and to ensure that the audience leaves promptly.

Indoors: Buildings designed for public assembly

125 Buildings designed for public assembly will have suitable and sufficient means of escape for their designed purpose. However adaptations, such as the provision of a stage, temporary stands, or a significant increase in the number of people to be accommodated, need to be taken into consideration and may require extra measures.

126 Where additions to the existing means of escape are needed, make sure that:

- exits are suitable and sufficient in size and number;
- exits are distributed so that people can turn their back on any fire which may occur;
- exits and exit routes are clearly indicated; and
- escape routes are adequately lit (see chapter on *Electrical installations and lighting*).

Indoors: Buildings not designed for public assembly

127 As it is unlikely that such venues were designed to accommodate large numbers of people, it is almost certain that additional means of escape will be required to accommodate a music event. Consult the fire and local authority at an early stage.

128 In deciding whether the means of escape are reasonable they will take into consideration:

- the occupant capacity of the building;
- the width and number of exits required;
- whether temporary stands and/or stages will be constructed within the building;
- exit and directional signs; and
- the normal and emergency lighting with which the building is provided.

Sports stadia

129 A sports stadium which has been issued with a general safety certificate under the Safety of Sports Grounds Act 1975 will be provided with adequate means of escape from the normal spectator areas. However, additional exits may be needed if the pitch area is to be occupied by the audience and/or by temporary structures, such as a stage or stands. If the stadium is designated under section 1 of the 1975 Act, a Special Safety Certificate is likely to be required for the event. Where such a certificate is required, apply to the relevant local authority as early as possible.

130 If a sports stadium is to be used which does not require certification under the legislation described in paragraph 129, or Part 3 of the Fire Safety and Safety of Places of Sport Act 1987, it is important to ensure that there are adequate means of escape from all areas. Consult the fire authority and local authority at an early stage. Further guidance in relation to the spectator and ancillary areas is given in the *Guide to safety at sports grounds* (see chapter on *Stadium events*).

Outdoor venues

131 Outdoor venues such as parks, fields and gardens of stately homes will normally have boundary fences at their perimeters. To provide means of escape which will allow for an orderly evacuation to take place, ensure that:

- the number and size of exits in the fences, etc, are sufficient for the number of people present and are distributed around the perimeter;
- exits and gateways are unlocked and staffed by stewards throughout the event; and
- all exits and gateways are clearly indicated by suitable signs which are illuminated if necessary.

132 At the planning stage, consult the fire authority and local authority about the proposals for means of escape.

General requirements

Marquees and large tents

133 Information concerning fire safety for temporary structures used for entertainment purposes, which includes marquees and large tents, can be found in the *Guide to fire precautions in existing places of entertainment and like premises.*

134 The Made-Up Textiles Association (MUTA) is also able to provide advice concerning the latest developments regarding doors and their fastenings, electrical and gas safety and escape routes (see *Useful addresses*). Information is also available in chapter 12 of the Institution of Structural Engineers document, *Temporary demountable structures: Guidance on design, procurement and use.*

Stairways

135 Any stairway, lobby, corridor or passageway, which forms part of the means of escape from the venue, should be of a uniform width and constructed and arranged so as to provide a safe escape for the people using it.

136 In general, stairways should be no less than 1.05 m wide. The aggregate capacity of stairways should be sufficient for the number of people likely to have to use them at the time of a fire. In this connection it will be necessary to consider the possibility of one stairway being inaccessible because of fire and the aggregate width should allow for this possible reduction.

137 Detailed guidance on exit capacity, related to evacuation time, is given in the *Guide to fire precautions in existing places of entertainment and like premises.*

138 Stairways wider than about 2.1 m should normally be divided into sections, each separated from the adjacent section by a handrail, so that each section measured between the handrails is not normally less than 1.05 m wide.

Ramps

139 Where ramps are used the:

- gradient should be constant and not broken by steps;
- maximum gradient for a ramp which is subject to heavy crowd flow should not exceed 1 in 12;
- ramp should have a non-slip surface and, as appropriate, have a guard rail and a handrail.

Note: Ramps installed for wheelchair users should conform to the British Standard BS 5810: 1979.

Exits

140 Every venue should be provided with exits that are sufficient for the number of people present in relation to their width, number and siting. Normally no exit should be less than 1.05 m wide. Full guidance on the calculation of exit widths and evacuation times for places of public assembly is given in the *Guide to fire precautions in existing places of entertainment and like premises* and for sports stadia in the *Guide to safety at sports grounds.*

Doors on escape routes

141 As a general principle, if a building is used for public assembly, a door used for means of escape should open in the direction of travel.

142 Also, the door should:

- not open across an escape route;
- be hung to open through not less than 90° and with a swing which is clear of any change of floor level;
- be provided with a vision panel if it is hung to swing both ways; and
- if protecting an escape route, be fire-resisting, fitted with smoke seals and be self-closing.

143 Any door which for structural reasons cannot be hung to open outwards should be locked in the fully open position at all times when the building or venue is occupied. The key should be removed to a safe place and the door should be clearly indicated with a sign bearing the words TO BE SECURED OPEN WHEN THE PREMISES ARE OCCUPIED. The notice should be provided on each side of the door in a position where it can be clearly seen whether the door is in the open or closed position.

Fastenings on doors and gates

144 Doors and gates which are final exits and all doors leading to such exits should be checked before the event starts to ensure that they are unlocked, or in circumstances where security devices are provided, can be easily and immediately opened from within, without the use of a key, by someone escaping. Security fastenings such as padlocks and chains should not, under any circumstances, be used when the venue is occupied; they should be placed on numbered hooks in a position which is not accessible to unauthorised people at all times when the building is occupied. All fastenings should be numbered to match the numbered hooks.

145 Where doors have to be kept fastened while people are present, they should be fastened only by pressure release devices such as panic bolts, panic latches or pressure pads which ensure that the door can be readily opened by pressure applied by people from within. Panic bolts, panic latches and pads should comply with BS EN 179: 1998 and BS EN 1125: 1997.

Self-closing devices for fire doors

146 It may be necessary for escape routes to be protected by fire-resisting construction and fire doors. All such doors, except those to cupboards and service ducts, should be fitted with effective self-closing devices to ensure the positive closure of the door. Rising butt hinges are not normally acceptable.

147 Fire doors to cupboards, service ducts and any vertical shafts linking floors should be either self-closing or kept locked shut when not in use and self-closing doors should be indicated by notices bearing the words FIRE DOOR KEEP SHUT. Doors to be kept locked should be indicated by notices bearing the words FIRE DOOR KEEP LOCKED.

148 All fire doors should be regularly checked to ensure that they are undamaged, swing freely, are closely fitted to frame and floor and that the self-closing device operates effectively.

Exit and directional signs

149 In an emergency, it is essential that all available exits are used. Clearly indicate all available exit routes so that members of the audience and workers are aware of all the routes to leave the venue in an emergency. In addition, the provision of well-sited signs and exit routes in full view of everyone present will give a feeling of security in an emergency.

150 All fire safety signs, notices and graphic symbols should conform to the Health and Safety (Safety Signs and Signals) Regulations 1996. In addition to the fire safety signs specified in the Regulations, signs which conform to BS 5499: Part 1: 1990 (amd 1995) and Part 3: 1990 will continue to satisfy the requirements of the Regulations.

151 Exit signs must take the form of a pictogram symbol but may be supplemented by text bearing the words EXIT or FIRE EXIT in conspicuous lettering. Any exit on an escape route should be clearly indicated by suitable exit signs positioned, wherever possible, immediately above the door or opening.

152 Where an exit cannot be seen or where people escaping might be in doubt as to the location of an exit, provide directional exit signs at suitable points along the escape route. Such signs should be sufficiently large, fixed in conspicuous positions, and wherever possible be positioned between 2 m and 2.5 m above the ground level.

153 Exit signs and signs incorporating supplementary directional arrows should be lit whenever people are present. Signs at outdoor events should be weatherproof and clearly visible above people as well as lit at night if necessary.

Normal lighting and emergency lighting

154 If used outside the hours of daylight, or in the absence of natural daylight, all parts of the venue to which the audience have access and all escape routes should be provided with normal lighting and emergency lighting (see chapter on *Electrical installations and lighting*).

Classification of fires

155 Fires are classified in accordance with BS EN 2: 1992 and are defined as follows:

Class A fires:	fires involving solid materials, usually of an organic nature, in which combustion normally takes place with the formation of glowing embers;
Class B fires:	fires involving liquids or liquefiable solids;
Class C fires:	fires involving gases;
Class D fires:	fires involving metals.

Class A fires

156 Class A fires are the most likely type of fire to occur in the majority of venues. Water, foam and multi-purpose powder are the effective media for extinguishing these fires. Water and foam are usually considered to be the most suitable media and the appropriate equipment are therefore hose reels, water-type extinguishers or extinguishers containing fluoroprotein foam (FP), aqueous film-forming foam (AFFF), or film-forming fluoroprotein foam (FFFP).

Class B fires

157 Where there is a risk of fire involving flammable liquid it will usually be appropriate to provide portable fire extinguishers of foam (including FP, AFFF and FFFP), carbon dioxide or powder types. Table 1 of clause 5.3 of BS 5306: Part 3 1985 gives guidance on the minimum scale of provision of various extinguishing media for dealing with a fire involving exposed surfaces of contained liquid.

158 Care should be taken when using carbon dioxide extinguishers as the fumes and products of combustion may be hazardous in confined spaces.

159 Dry powder extinguishers can have an effect on visibility and breathing if used in a crowd of people or in a confined space. Incorrect use could possibly cause a degree of panic.

Class C fires

160 No special extinguishers are made for dealing with fires involving gases because the only effective action against such fires is to stop the flow of gas by closing the valve or plugging the leak. There would be a risk of an explosion if a fire involving escaping gas was extinguished before the supply was cut off.

Class D fires

161 None of the extinguishing media referred to in the preceding paragraphs will deal effectively with a fire involving metals such as aluminium, magnesium, sodium or potassium, although there is a special powder which is capable of controlling some Class D fires. However, only specially trained workers should tackle such fires.

Fire-fighting equipment

162 The following paragraphs give advice on fire-fighting equipment for use in the early stages of a fire before the arrival of the fire brigade. Some venues designed for public assembly may have a fire suppression system, eg a sprinkler system, but generally portable or hand-held fire-fighting equipment, ie extinguishers, hose reels and fire blankets will be sufficient.

163 All venues should be provided with appropriate portable or hand-held fire-fighting equipment and this provision should be determined at the planning stage in consultation with the local authority and fire authority.

Fire extinguishers

164 If portable fire extinguishers are installed, they should conform to BS EN 3: 1996 and be colour coded in accordance with BS 7863: 1996 and BS 5306: 1985.

Hose reels

165 If hose reels are installed they should be located where they are conspicuous and always accessible. The hose should comply with Type 1 hose specified in BS 3169: 1986 and hose reel installations should conform with BS 5306: Part 1 1976 and BS EN 671: Part 1 1995.

Fire blankets

166 Fire blankets are suitable for some types of fire. They are classified in BS 6575: 1985 and are described as follows.

- Light duty: these are suitable for dealing with small fires in containers of cooking fat or oils and fires in clothing.
- Heavy duty: these are for industrial use where there is a need for the blanket to resist penetration by molten materials.

Fire involving electrical equipment

167 The use of water-type fire extinguishers where there is any electrical supply is dangerous. Extinguishers provided specifically for the protection of electrical risks should be of the dry powder or carbon dioxide type. While some extinguishers containing aqueous solutions such as AFFF may meet the requirements of the electrical conductivity test of BS EN 3: Part 2 1996, they may not sufficiently reduce the danger of conductivity along wetted surfaces such as the floor. Consequently, such extinguishers should not be provided specifically for the protection of electrical risks.

Fire-fighting equipment provision

Indoors: Buildings designed for public assembly

168 Usually, the scale of provision required in connection with the normal use of the building will be adequate. However, if additional facilities are to be provided, eg a stage, concessions on a pitch, changing rooms, etc, there may be a need for additional equipment.

Indoors: Buildings not designed for public assembly

169 These venues cause the greatest concern as existing provisions may be minimal. However, there may be some provision (eg hose reels in a warehouse) and provided that the maintenance is satisfactory, this should be taken into account. In deciding what fire-fighting equipment is appropriate, consider both the structure and the contents of the building including the scale of both. The general principle is that no one should have to travel more than 30 m from the site of a fire to reach an extinguisher. Position extinguishers on exit routes near to exits.

Outdoor venues

170 The provision of fire-fighting equipment for outdoor venues will vary according to the local conditions and what is brought onto the site. There will need to be equipment for tackling fires in vegetation, vehicles and marquees. The best arrangement is to provide well indicated fire points as follows:

- where water standpipes are provided on site and there is a water supply of sufficient pressure and flow to project a jet of water approximately 5 m from nozzle, fire points consisting of a standpipe together with a reel of small diameter hose of no less than 30 m in length should be provided. Provide the hose with the means of connection to the water standpipe (preferably a screw thread). The hose should end in a small hand-control nozzle. Keep hoses in a box painted red and marked 'HOSE REEL';
- where standpipes are not provided or the water pressure or flow is not sufficient, provide each fire point with either a water tank at least 25 L in capacity fitted with a hinged cover, two buckets and one hand pump or bucket pump; or a suitable number of water-type fire extinguishers (not less than two No 13A rated extinguishers).

171 Arrangements may need to be made to protect fire-fighting equipment located outdoors from the effects of frost, vandalism and theft. Fire points should have prominent signs. Further advice should be sought from the fire authority or local authority.

Special risks

172 In addition, provide portable fire-fighting equipment for special risks in accordance with the following scale:

- Stage exceeding 56 m^2: Hydraulic hose reels or two water-type extinguishers (rating 13A), on each side of the stage, and one light-duty fire blanket (see paragraph 167 regarding electrical equipment);
- Stage not exceeding 56 m^2: One water-type extinguisher (rating 13A), on each side of the stage, and one light-duty fire blanket (see paragraph 167 regarding electrical equipment);
- Dressing rooms: In every block of four dressing rooms a minimum of one water-type extinguisher (rating 13A) and one light-duty fire blanket;
- Scenery store, stage basement, property store and band room: Water-type extinguisher (rating 13A) in each risk area, or an appropriate extinguisher where water is unsuitable for the fire risk presented;
- Electrical intake rooms, battery rooms, stage switchboards and electrical equipment: Carbon dioxide extinguisher or one dry-powder extinguisher (minimum rating 21B);
- Boiler rooms - solid fuel fired: Water-type extinguisher (rating 13A);
- Boiler rooms - oil fired: One dry-powder or foam extinguisher (rating 34B);
- Portable generators (power supply): Carbon dioxide extinguisher, or one dry-powder extinguisher;
- Mobile concessions: One dry-powder extinguisher (rating 21B) and one light-duty fire blanket. (see chapter on *Food, drink and water*)

Means of giving warning in case of fire

173 The following paragraphs give general advice on the means for giving warning in the event of fire. More detailed advice may be obtained from the *Guide to fire precautions in existing places of entertainment and like premises* or from the fire authority and local authority.

Fire-warning systems

174 The purpose of a fire-warning system is to provide information to stewards and everyone present so that all can be safely evacuated before escape routes become impassable through fire, heat or smoke. The means for giving warning should be suitable for the particular venue, taking into account its size and layout and the number of people likely to be present.

175 Fire-warning systems should generally comply with BS 5839: Part 1 1998. The Health and Safety (Signs and Signals) Regulations 1996 requires that a sign or signal that needs a power supply to operate should also have a back-up power supply. Existing systems designed or installed to an earlier standard may be acceptable subject to satisfactory testing, electrical certification and approval by the local authority (see chapter on *Communication* for further advice on emergency public announcements).

Indoors: Buildings designed for public assembly

176 A venue which has an existing entertainment licence for music events will have an approved means for giving warning in case of fire. However, it will be necessary for the fire authority to be consulted at an early stage to ensure that the system is appropriate.

Indoors: Buildings not designed for public assembly

177 Buildings not designed for public assembly such as warehouses, aircraft hangars, agricultural buildings, etc, may have a warning system which is unsuitable for a music event or no fire-warning system at all. It will therefore be necessary to either modify the existing system to use the building for the event or provide a temporary warning system.

178 If a temporary warning system is installed (and this may be the more appropriate action to take), the provision of a radio-transmission system has a number of advantages as it will not require the laying of electrical wiring or modifications to a building. Static call-points can also be replaced by mobile call-points carried by stewards so that the alarm can be raised instantly at the point of discovery of any fire. It is, however, still necessary for any system to comply with the general principles of BS 5839: Part 1 1998 and reference should also be made to BS EN 60849: 1998. The fire authority and local authority should be consulted as to the suitability of the system for the venue.

179 For some buildings not designed for public entertainment, an alarm system incorporating automatic fire detection may be required, particularly in circumstances where a fire could reach serious proportions before discovery.

Outdoor events

180 Although there is less likelihood of people becoming trapped by fire when the event is staged outdoors it will still be necessary to provide a fire-warning system for temporary and moveable structures such as marquees. Campsites should have fire watchtowers and campers should be provided with fire safety advice.

Fire safety advice on curtains, drapes and other materials

181 The use of curtains, drapes, and temporary decorations could affect the safe use of the means of escape. Any proposal to use combustible decorative materials should be notified to the fire authority and local authority in writing and should be accompanied by full details, including samples (not less than 1 x 0.5 m) of the material proposed to be used. Where a building is already being used for public assembly the use of these materials will probably have been approved.

Curtains and drapes

182 All curtains and drapes should be of durable or inherently flame retardant fabric and should conform with BS 5867: Part 2 1980 (amd 1993) Fabric Type B when tested in accordance with BS 5438: 1998. Non-durable fabric is acceptable provided that it conforms to BS 5867: Part 2 1980 (amd 1993) Fabric Type B and is accompanied by a test certificate.

183 Where doubt exists about the flame retardancy of a material, obtain a test certificate to show compliance with the appropriate standard. Tests should be conducted by an approved laboratory under the Department of Trade and Industry's National Measurement Accreditation Service (NAMAS) scheme or similar approved test laboratory.

184 Curtains across exit doors present an additional problem and should be arranged so as not to trail on the floor. They should open from the centre and should only be permitted where stewards are present nearby to open the curtain in the event of an emergency.

Artificial and dried foliage

185 All artificial and dried foliage used for decorative purposes in audience areas should be flame retardant. As the flame-retardant treatment can be adversely affected by contact with moisture, periodic re-treatment may be necessary to maintain its effectiveness. Re-treatment may also be necessary to maintain the appearance of the foliage.

186 There are no laboratory test methods for assessing the flame-retardant properties of dried or artificial foliage. However, it is recommended that these and similar items should be subject to ignition tests using small flaming sources comparable to those used for testing drapes and curtaining.

187 As it is difficult to totally inhibit the production of flaming molten droplets or debris from the solid plastic parts of artificial foliage such as branches and stems, the fire/local authority may limit the amount of material used and prohibit use in some locations.

Chapter 4

Major incident planning (emergency planning)

188 The consequences of a major incident at a music event could be catastrophic and it is necessary to plan for such an occurrence. A major incident will normally require a multi-agency approach in which the event organiser, police, NHS including the ambulance service, fire authority, local authority, local emergency planning officer, stewards and first aiders may play a part. It is therefore important that there is a clear demarcation of duties and that responsibilities are agreed and understood at the event planning stage. Agreed procedures should be issued in writing to all relevant parties.

189 Procedures to deal with serious and imminent danger in the workplace including evacuation are a requirement of the Management of Health and Safety at Work Regulations 1992 (Management Regulations) (see chapter on *Health and safety responsibilities*).

Definitions

190 The agreed definitions of a major incident can be found in paragraph 192. Minor emergencies or incidents that do not require the intervention of the emergency services, NHS or local authority will need to be dealt with by developing suitable contingency plans. It is important to appreciate that a minor incident could have the potential to develop into a major incident if not properly planned for and managed. Event organisers should therefore develop contingency plans to deal with minor incidents along with their major incident plans. Major incident plans should be developed in conjunction with the emergency services.

191 It is of fundamental importance to identify in your plans precisely what needs to be done and agree the situations in which it will be necessary to hand co-ordination of an incident over to the police. This could be before any actual major incident has taken place if it is thought that a handover might prevent an incident from developing. It is also important to agree with the emergency services the procedures for declaring a major incident and who declares it.

Defining a major incident

192 A major incident is any emergency that requires the implementation of special arrangements by one or more of the emergency services, the NHS or the local authority for:

- the initial treatment, rescue, and transport of a large number of casualties;
- the involvement either directly or indirectly of large numbers of people;
- the handling of a large number of enquiries likely to be generated both from the public and the news media, usually to the police;
- the need for the large scale combined resources of two or more of the emergency services;
- the mobilisation and organisation of the emergency services and supporting organisations, eg local authority, to cater for the threat of death, serious injury or homelessness to a large number of people.

Further information on major incident planning can be found in the Home Office publication *Dealing with disaster.*

Planning

193 The event risk assessment will be a good starting point for any major incident plan. This will help you focus on areas that will need to be considered. Areas include:

- the type of event, nature of performers, time of day and duration;
- audience profile including age, previous or expected behaviour, special needs, etc;
- existence or absence of seating;
- geography of the location and venue;
- topography;
- fire/explosion;
- terrorism;
- structural failure;
- crowd surge/collapse;
- disorder;
- lighting or power failure;
- weather, eg excessive heat/cold/rain;
- off-site hazards, eg industrial plant;
- safety equipment failure such as CCTV and PA system;
- delayed start, curtailment or abandonment of the event.

Preparation of major incident plans

194 Consider the following matters when preparing your major incident plan:

- identification of key decision-making workers;
- stopping the event;
- identification of emergency routes and access for the emergency services;
- people with special needs;
- identification of holding areas for performers, workers and the audience;
- details of the script of coded messages to alert and 'stand down' stewards;

- alerting procedures;
- public warning mechanisms;
- evacuation and containment measures and procedures;
- details of the script of PA announcements to the audience;
- identification of rendezvous points for emergency services;
- identification of ambulance loading points and triage areas;
- location of hospitals in the area prepared for major incidents and traffic routes secured to such hospitals;
- details of a temporary mortuary facility;
- an outline of the roles of those involved including, contact list and methods to alert them;
- details of emergency equipment location and availability; and
- documentation and message pads.

195 Your plan should provide a flexible response whatever the incident, environment or available resources at the time. It may be necessary to prepare variations of the plan to deal with specific issues. Your plan should also build on routine arrangements and integrate them into the existing working procedures on site.

196 Experience has shown that a multi-agency approach to all planning will share the ownership of problems and lead to effective solutions. This approach can be termed integrated emergency management. A planning team should be created from people and agencies who will be required to respond to any emergency or major incident.

197 To be effective, the major incident planning team should not be too large. It may be useful to have a number of specialist subgroups. Each organisation, eg police, fire brigade, first-aid provider, etc, concerned with the event should give a clear undertaking as to their role and committed resources if a major incident happens. This will be in the form of a statement of intent.

198 The person leading the planning team must be competent to do so and have a broad appreciation of the issues. This person does not necessarily have to be the event organiser or one of their workers. However, they will be accountable for the plan's effectiveness and for the person chosen to lead the team. The event safety co-ordinator should be involved in the planning process. Keeping and retaining records of meetings and decisions is very important.

199 The plan should be easily understood and without jargon. Instructions, particularly in respect of action to be taken, must be specific so that a named person/role/rank will carry out a specific function. A glossary of terms may assist. Much time can be saved if the layout of the plan allows for simple and quick updating. Revised copies should be easily identifiable from a date/numbering system.

200 Off-site implications will form an important part of the plan. Traffic issues will include emergency access and exits, as well as readiness for an off-site incident occurring with consequences for the event. This could include a coach crash or large numbers of visitors stranded. Where a venue is close to county or other administrative boundaries, liaison may be required by the emergency planning officers of the local authority and the ability to provide mutual aid determined. Consult the local authority emergency planning officer in relation to the existing local authority emergency plans and give a copy of your event major incident plan to the local authority emergency planning officer.

201 Detailed, gridded site plans containing pertinent geographic and topographic features will be of great value during planning and in the event of a major incident. They will be particularly useful when calculating normal and emergency pedestrian flow.

Training, exercising and testing

202 Think about testing the plan to check its effectiveness and the competence of the individuals and teams who will operate it. Methods can include simulation exercises or table-top exercises. Exercises need not be full scale and may be designed to test only one element of the plan at a time. Debriefing following an exercise is particularly constructive and will dispel misunderstandings that may have arisen and strengthen future working relationships.

203 Once the plan has been agreed, each organisation must ensure that the people responsible for putting the plan into practice are fully briefed. By doing so, problems can be prevented in the first instance, but if one occurs, properly briefed workers can stop a situation deteriorating. Communication exercises are strongly recommended before the event. The training of stewards is also an essential safety element. Stewards and others likely to have an emergency role, must be issued with written details of their duties, major incident procedures and a gridded site plan. Brief relevant people connected with the event, including concessionaires and those supplying other services who could be in a position to provide important assistance.

204 A major barrier to effective briefing is the transient nature of stewarding and the shift working by the emergency services. This situation can be made more difficult when additional workers are hurriedly brought in. Methods of informing workers in these circumstances can include individual, team or group presentations, written instructions and training videos.

Emergency service and local authority responsibilities

205 Once a major incident has been declared the police will co-ordinate and facilitate the 'on-' and 'off-site' response. However, in the case of a fire, the fire brigade will be responsible for dealing with an on-site response. The NHS ambulance service will initiate co-ordination of the overall medical response at the scene, nominating and alerting receiving hospitals, distributing casualties, providing emergency transport, communications and liaison with the other agencies. Local authorities are able to provide a range of services in case there is a major incident. Services may include reception centres, temporary emergency accommodation, feeding and access to a wide range of special equipment.

Cordons

206 In the event of a major incident, cordons may be needed. Discuss with the police, fire brigade and ambulance service how this would be carried out on site. Place cordons according to the circumstances. They may need to be moved during the course of the incident.

Major incident management structure

207 Across the UK there has been widespread adoption within the emergency services and local authorities of a three-tier management structure for dealing with major incidents. This recognises that in very serious situations there may be a need to co-ordinate the handling of an incident at operational, tactical and strategic levels. Many event organisers already use this model and you may wish to consult the emergency services in your area to see whether it is appropriate for you to adopt it. (In some areas, these different levels of management are referred to as bronze, silver and gold respectively, but in others the latter terms are not used.)

208 Regardless of the terminology, this mutually agreed system offers a simple management structure which eases co-ordination between responding agencies. Each agency is responsible for putting this structure into practice for its own activities.

209 The 'operational level' involves the managers closest to the incident who are managing

deployment and execution of tasks within a geographical sector or specific function. There may be any number of operational managers and most incidents will be handled at this level, only moving on to the next level should the nature of the incident make this necessary.

210 'Tactical managers' are responsible for formulating the tactics which will be pursued by their organisation. Tactical managers normally attend the scene, but when more that one organisation operates at this level there must be consultation between them. Their tasks involve overall co-ordination, general management of the incident and deciding how resources will be allocated.

211 A higher 'strategic level' of management may be needed for the most serious incidents. Strategic management is best achieved away from the scene. Strategic managers are responsible for formulating for their own organisation the policy framework within which their tactical commanders will work, prioritising organisation demands as a whole. Where more that one organisation needs to operate at this level a 'strategic co-ordinating group', generally chaired by the police, will be formed to ensure that the strategies of the different agencies involved are compatible.

Incident control rooms (the co-ordinating group and location)

212 Other than at small events, it is essential that on-site accommodation is set aside as a designated emergency liaison centre or incident control centre. While the event is running, make sure this on-site facility is staffed continuously. Consider the location of this incident control centre in the overall venue and site design (see chapter on *Communication*).

Emergency service control vehicles

213 If there is a major incident, the emergency services are likely to dispatch their command and control vehicles to the scene. Clearly there will be benefits if their vehicles can be situated near to the emergency liaison centre and so consider this factor in your overall venue or site design.

Communication

214 Advice on communications and emergency public announcements can be found in the chapter on *Communication*.

Media management

215 The moment a major incident develops the media will be making enquiries. Some may already be present covering the event while others will quickly arrive at the scene. Plan to provide them with an accurate and credible response by developing a suitable strategy. It is important that all parties concerned with the event appreciate the media's need to gather information. Consider appointing a chief press officer and identify a media rendezvous point to help with media liaison. In the event of a major incident, the police media manger is responsible for the co-ordination of the response to the media.

Scene and evidence preservation

216 Any major incident is likely to result in an inquiry that may lead to criminal and civil proceedings. The police, fire brigade, health and safety inspectors and local authority officers carry out evidence gathering and investigations. In the first instance it will be the responsibility of the police to ensure that the scene and any other evidence is preserved. Obviously, this action will not interfere with saving of life. Make sure that you are clear as to which officers and inspectors will need access to information to carry out any necessary investigations.

Voluntary agencies

217 Many voluntary agencies can provide high-quality aid at incidents and if they are available at your event consider involving them in your emergency planning (see chapter on *Information and welfare*).

Some specific scenarios

Cancellation of an event

218 If an event needs to be cancelled after the audience has arrived, or a performance has begun, stopped and not re-started, there will be a wide range of issues to be managed. This will be so even if there has not been an actual major incident. Property may have been lost or abandoned and people stranded. There may also be an expectation for compensation or the re-issuing of tickets. Think about preparing statements which can be given to the audience together with a press release to the public.

Stopping and starting an event

219 Once the music event has begun, unscheduled stopping of the event could present serious hazards. Any decision to do so must be taken after careful consideration and consultation with the major incident planning team. Likewise, deciding whether or not, and when to evacuate the audience will require fine judgement. Both unscheduled stopping and evacuation are scenarios that must be pre-planned and as far as practicable, tested and rehearsed. The major incident plan must state who it is that makes the decision to stop or start the event.

Bomb threats

220 If a telephone bomb threat is received details of the call must be recorded as accurately as possible. (The police are able to provide guidance on this.) It is essential that the information is immediately passed to the police for evaluation and response.

221 The police will advise on the validity of a threat. Generally, any decision to evacuate or move people will rest with the event organiser. The exception is where a device is found or where police have received specific information. In these circumstances the police may initiate action and the directions of the senior police officer present must be complied with. If a bomb is a real threat, care must taken to be alert for secondary devices. These might be aimed at the emergency services or the moved/evacuated audience.

Chapter 5

Communication

222 Effective communication is of prime importance if an event is to run smoothly and safely. Communication requirements of all the organisations involved in the event (assessed individually or jointly) need to be examined thoroughly. This includes examining the general and operational management of the event, handling routine health, safety and welfare information and communicating effectively in the event of a major incident.

223 This chapter explores key communication issues from two main perspectives:

- inter-professional communication;
- public information and communication.

Inter-professional communication

Communication during the event planning phase

224 The communication network during this phase is wide and involves a range of communication activities and information requirements:

- intelligence gathering about the event characteristics, etc;
- liaison meetings;
- seeking appropriate licences;
- preparation of detailed plans for arrangements on and off site;
- commercial arrangements - ticketing policy, publicity, contracts, etc.

225 Everyone involved in the planning of an event will need to keep proper records of decisions and ensure that relevant information is communicated to others. It is particularly important that statement of intent documents are clear and unambiguous in their definition of roles and the responsibilities of different agencies and individuals.

Preparation of key support documentation

226 Unambiguous use of language is crucial in providing a clear and reliable communication framework. Avoid jargon and acronyms wherever possible. Where they are necessary, it is worth including a glossary of terms within the main planning documents.

227 Agree special terminology to be used by people preparing plans, documents and communication procedures in relation to:

- naming different control points and control workers;
- labelling different types of rendezvous and collection points;
- providing unique reference labels for key locations within and around the venue;
- clear naming conventions for categories of people involved on site;
- compatible terminology for assessing risks and grading levels of urgency;
- clear contact protocols for establishing communication.

228 Wherever possible, plans should say who does what, not just what is to be done. For example, 'the incident control room must be informed', is not as helpful as, 'the duty officer must inform the incident control room'.

229 (The Plain English Campaign produces much helpful guidance about many aspects of written communication, see *Writing plain English.*)

230 Relevant maps and site plans are crucial. Visual data should show key routes for vehicles and people, and restrictions on access. A gridded site plan for the venue and its immediate surrounds is recommended. Discrepancies can result in delayed responses, misdirected resources and communication channels being unnecessarily blocked with requests for clarification and attempts to sort out the confusion.

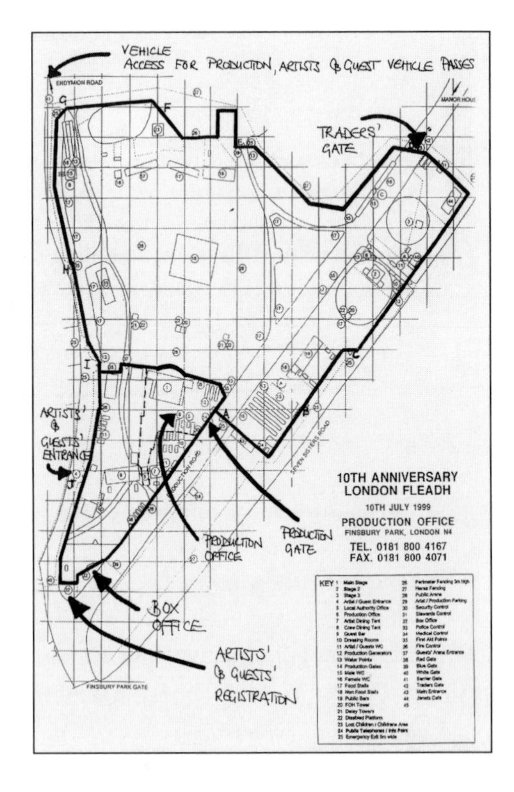

231 Pay attention to labelling features and functions consistently in different documents. If a feature occurs more than once (eg if there are several first-aid points) each should have a unique reference. Consult before altering plans so that the consequences of changes can be considered.

232 Consider appointing a network co-ordinator as a single point-of-contact. Such a person would receive, collate, cross-check and spread information about radio-channel frequencies, call signs, phone lines, alert cascades, camera points, siting of control equipment, contact lists, etc.

233 Ensure that major incident plans are compatible with emergency plans drawn up by local authority or emergency services. Make sure that relevant information is easily available to people in control rooms at remote locations.

234 Many other types of documents (technical diagrams, safety certificates, licences, approvals, minutes of meetings, etc) will figure in the overall communication processes, reference, guidance, authorisation or approval. Keep all documents up to date and inform people involved in the planning process of any changes immediately.

A framework for handling the event

235 There is a need for a framework that allows effective communication:

- within each organisation (individual emergency services, event organiser, stewarding organisations, local authority);
- between different agencies (police to fire, stewarding organisation to ambulance, venue operator to police, etc).

236 Such a framework involves both on-site and off-site links so that:

- organisations which need to respond to events on site can be easily contacted;
- appropriate organisations can be informed of events on site that may have off-site repercussions.

Communication controls in incident control rooms

237 Consider the following matters in relation to your event.

- Power supplies for communication equipment should be independent of production power supplies and with independent back-up facilities. Test power supplies for their ability to provide continuity of communication when switching over to auxiliary power. It is important that the back-up supply is adequate. In a major incident, this supply may need to last beyond a scheduled event finish time.
- Ensure that incident control rooms or 'units' have a clear view over as much of the event as possible, are easily accessible and have adequate space for equipment and for workers to operate effectively.
- Links should be available to allow communication between key personnel.
- Arrangements should ensure that communication is possible between incident control rooms and critical locations and activities.
- Co-locate communication controls for stewarding/security, emergency services, local authority and first-aid providers wherever practicable.
- Staffing should be sufficient to allow for periodic policy and review meetings between personnel from different organisations.
- Radio controllers must have the option to stop 'talk-through' facilities if an urgent situation develops.
- Route all cabling and wiring through areas of low risk from fire or other damage.
- Arrange for maintenance workers to be on hand to carry out any necessary repairs or adjustments.
- Appropriate levels of soundproofing are essential and where appropriate, provide workers with headsets to cut out interference from noise within a busy control room.
- Provide key items of documentation and stationery in all control rooms: site plans, key contact details, alerting cascades, message pads, log sheets, etc.
- Display frequently-used information clearly (site plans, key contacts, etc) and make sure facilities such as wipeable boards or flip charts are available for writing up incident-specific information as it arises.
- The need to maintain and operate emergency communications from an alternative site.

The *Guide to safety at sports grounds* contains much practical guidance on communications.

Off-site links

238 Provide details of the event in the control rooms of each of the emergency services and ensure that communication lines, whether by radio or telephone, to the local headquarters of all emergency services are available at all times so that emergency calls can be made instantly.

239 Consider arrangements for communicating with organisations that are affected off site by movements of large numbers of people, eg traffic police, British Transport police, transport providers, etc. This is particularly relevant when unforeseen events (such as curtailment of an event) could have significant knock-on effects at locations remote from the event itself.

Radio communication

240 Depending on the size of the event, there can be many radio sets and networks operating simultaneously on site. Inform contractors of the frequencies that are available, before they hire radios. The network co-ordinator can collect information on all proposed frequencies and consult with the Radio-communications Agency and local emergency services.

241 Each organisation requiring radio communication will need to consider what operational channels are necessary for identified functions or areas. In addition, emergency services will have to consider the need for command channels at large events.

242 Radio is an important medium for general operational requirements and a prime medium for responding to emergencies. Pre-event checks are therefore essential. Carry out full perimeter tests to ensure coverage is adequate. At an outdoor site, appropriate positioning of masts, antennae and repeaters may require research and testing. The erection of temporary structures can have a significant impact on radio coverage and corrective measures may be necessary.

243 Background noise is obviously a huge problem when working in any large audience and when there is loud music. The issuing of full ear-defending headsets should be considered for key workers in high-noise areas and remember to fully charge all batteries at the start of the event. Adequate numbers of spare batteries and charging facilities are essential. The risk of wearing such equipment will have to be assessed in relation to any physical-injury hazards associated with the kind of tasks they have to carry out.

Telephone equipment

244 Provide external lines for immediate telephone contact between the venue control points and emergency services control rooms off site. Do not use external telephone lines designated for emergency use for other communication.

245 Field telephone networks (or internal telephone networks in a venue such as a sports stadium or arena) provide vital links between on-site communication controls and other key points around the venue. Cell phones are widely used and provide extra communication options. However, they should not be relied upon for important links and especially not used for emergency communication.

Closed circuit television (CCTV)

246 A valuable safety and security tool is CCTV which can assist crowd management. Certain fundamental questions are worth considering in the event planning stage:

- will the use of CCTV make the event safer?
- where should cameras be located?
- will there be sufficient light?
- who should have control over them?
- who should have viewing access?

247 The CCTV images can greatly enhance the potential to identify problems in a crowd resulting from surges, sways, excessive densities or public disorder.

Communication procedures

248 There must be a clear framework of information flow procedures - people need to know who should inform whom of what, when, and by what means.

249 Prime concerns are:

- tight radio discipline with proper use of call signs and contact protocols;
- making the purpose/function of a message clear (is it a question, warning, request for action, command, prohibition, etc);
- concise and precise information;
- cross-checking that messages have been received and interpreted correctly;
- relaying message content clearly and unambiguously;
- keeping accurate records of communication activity;
- keeping accurate logs of decisions and actions.

Message delivery and acknowledgement

250 Workers must be aware of the possible consequences if messages are not properly communicated and understood. There will be marked differences in levels of local knowledge among workers at and around the event and therefore procedures for acknowledging or reading back messages should be introduced.

Situation reports

251 Develop procedures for providing information from the scene of an incident or emergency. Note that:

- a practised format helps the person providing information to include necessary details for an appropriate response;
- a familiar communication pattern helps people receiving information to anticipate and recognise items; this assists the receiver to note the information ready for subsequent use or relay.

252 A situation report format must work equally well for any type of incident. It is particularly important to include the following items of information in such a report.

Identification:	call signs, names of calling and called parties;
Location:	exact details of where the incident is;
Incident:	precise details of what is involved;
Requirements:	details of services, equipment and agencies required.

253 Obtain acknowledgement that these key items have been received and understood before giving further details. If more information is available, further items that are particularly important in second or further transmissions are:

Warnings:	details of any hazards (present or potential);
Access:	any details about what might affect access to the scene, or advice on the quickest access route;
Casualties:	any details known about injured or sick people;
Control point:	details of who to contact and where, for more information from the scene;
Other information:	any other relevant information.

Alerting procedures

254 Different levels of urgent response may be necessary, for anything from serious injury or sickness of a single person, to a large-scale response to a major incident that requires full and urgent evacuation. Accurate information is vital to appropriate decision making. Since each link in a communication chain is a potential source of misunderstanding or breakdown, alerting chains between informants, decision makers and response providers should be as short as possible.

255 Alarms or threats will have to be evaluated on their merits, but in those circumstances where the first response is to investigate further, specific instructions or coded announcements should reach all those who have an emergency role. All workers involved should 'stand by' at designated emergency positions and await further information.

Regular updating

256 Good use of communication channels that allow easy access to information is essential. Whiteboard logs are a simple example of this. They are effective 'broadcasting' mechanisms - providing ready access to information and reducing verbal and hand-written exchanges. Much thought has to be given, therefore, to anticipating the information requirements of others.

Record keeping

257 Keeping records and logging information throughout the course of an event is a key activity. Logs must show key events and actions in sequence and are a valuable tool for keeping workers informed of the progress of any incident.

Training, briefing and preparation

258 All organisations have a responsibility for training their workers appropriately, covering everything from using appropriate radio discipline to keeping a decision log. There must be proper briefings for all workers about their duties for the event. This includes briefing workers offsite who need to be aware of special arrangements for an event, eg those in incident control rooms.

Public information and communication

Types of information

259 The information requirements of the audience range from performance details, ticketing arrangements, travel options, recommended routes, location of facilities, venue layout and welfare information right through to urgent contact messages or emergency instructions. Anticipating public information needs has an important bearing upon welfare and safety. Consider what information the audience will require if the event is cancelled or curtailed and how to provide that information. Well-informed people are less likely to be frustrated, aggressive or obstructive. Advance information on how to get to the venue, where to go on arrival or what will or not be allowed, all reduce frustration and irritation. If there is a need to communicate rules and restrictions, people are more likely to comply if they are aware of the reasons behind them.

Communication channels

260 Communication channels include:

- publicity material and tickets;
- media (press, radio, TV);
- route-marking;
- signs;
- notices, information displays;
- screens, scoreboards;
- face-to-face contact;
- emergency public announcements;
- PA systems.

Alarms

261 Audible alarms are useful alerting devices but convey little information. The activation of an audible alarm will most often need to be followed by an explanation about what to do, or simply information that it has been a false alarm.

PA systems

262 PA systems are a vital channel of communication with the audience. Output should be clear and intelligible for everyone of normal hearing in all parts of the venue, including people in the immediate surrounds. Ensure that the PA announcer has a good view over as much of the venue as possible and good communication links with control points. In the event of a major incident, override facilities must allow announcements to be made over the PA system without interference from other sound sources. Agree the circumstances in which this will happen in your major incident plans. The PA system should be fully tested before the event. It should have a back-up power supply that allows it to continue to operate at full load in the event of an emergency.

263 The availability of vehicle-mounted PA systems around the venue has also proved important when there is a need to communicate with people arriving at, or leaving, the venue. In the event of an evacuation there will be a need to communicate with the people outside the venue. The power requirements for these systems will need to be planned. Further useful guidance on PA systems can be found in the *Guide to safety at sports grounds* and the Football Stadia Advisory Design Council (FSADC) publication *Stadium public address systems*.

Screens, scoreboards

264 Video screens and scoreboards are a useful communication channel for putting out public announcements. They can provide information without interruption to a performance. For urgent public announcements, however, they can reinforce the message and give information to those who have hearing difficulties.

Loud hailers

265 Provide loud hailers at strategic points in the venue for use by stewards and police for urgent communication and as a back-up in case the PA system fails. Train workers how to use them and where they are located. Keep batteries fully charged.

Staff (face-to-face contact)

266 Direct contact between personnel and the public is obviously a vital communication channel, particularly in the safety chain. Approachable and helpful stewards have a particularly important role in creating a positive relationship with the audience. Their role in giving people clear and concise directions and assistance in an emergency can be a vital one.

267 Staff with any safety role should be easily identifiable by tabards or other high-visibility items of clothing. These allow the public to seek them out as a source of assistance and to recognise their authority when appropriate. If people are being directed along a route of safety, stewards in high-visibility clothing can help indicate the way much more clearly.

268 When problems are being dealt with, high-visibility clothing also helps colleagues, supervisors or CCTV controllers to pick them out and spot when they may be in difficulty or need support. In some cases, for workers who do not normally need to be visually conspicuous but may need to be identifiable for certain contingencies, reversible jackets that are high visibility on one side are worth considering.

Emergency public announcements

269 Where there is known danger, early warning is essential. Many estimates of crowd evacuation times only calculate the time between a crowd starting to move and the time the area has been cleared. However, in many situations, the time between first requesting people to evacuate an area and when they start to comply is a significant factor in the overall evacuation. Persuasion time must be added to movement time.

270 One reason often given for not indicating the cause of an evacuation is the fear of causing panic. Lessons learned from previous incidents and experimental research indicate that complacency can be a much greater danger. If panic occurs, it is usually only when people's escape time is so compressed that they feel they no longer have the chance to escape. Delayed communication has generally been found to be a far greater danger. Accurate and timely information is therefore essential.

271 People will respond better if the information comes from a source that is recognised as having authority or from someone the listeners respect. There is therefore a choice between using an authority or an empathy figure. If the latter seems to be the best option, seek the collaboration of the performers and brief them accordingly at the start of the event.

272 The major incident plan must indicate who has the authority to decide that an emergency announcement is necessary, who should make it and under what circumstances. Pre-plan the wording of announcements or text messages as far as possible. Agree messages to be made, through consultation between organisations with a safety role.

273 Coded announcements may be needed to alert workers so that they can take up emergency positions. Once they are ready, information can be given to the audience. The audience should be attracted by loud and distinct signals (such as two-tone chimes). Succeed in attracting their attention before any information is given. Keep language simple, sentences short and positive whenever possible.

274 Instructions should tie in with route marking, signs and other visual cues. Announcements need to make clear and specific location references wherever practicable (if possible, avoid relative references that can be interpreted differently such as 'away from the front' or 'further back'). In serious situations, an integrated approach would combine spoken announcements, text displays and directions from stewards in high-visibility clothing. Different people will respond to cues differently and this approach also recognises that some channels are not available to people with hearing or sight impairments. Reinforcement and repetition is needed to keep people on course.

275 Key elements should be repeated: what action is required, the nature of the incident, where to go, what to do on arrival. People should be told that it will be repeated as they proceed along their route. If not, they may be afraid of missing information and impede flows by stopping to check, seeking an authority figure to ask, or waiting for a further announcement.

276 Keep people up to date with the situation, even if no changes have occurred. Telling people that there is no change is still information, pre-empts individual queries and removes pressure from workers. Broadcasts are much easier and much more efficient than individual responses.

277 Even when people have left a venue they will need to know when and if an event will be restarted, whether and when they can collect vehicles, and what to do if the event is postponed, rescheduled or cancelled.

278 Summary of key points for emergency announcements.

- Early warning/timely information is essential.
- Persuasion time must be added to movement time.
- Clarity and quality of announcement delivery are crucial.
- Consideration should be given to whether an audience will respond better to an empathy figure making certain announcements.
- Live, directive messages relating to the circumstances are more effective than those that are pre-recorded.
- Reasons for messages (the nature of problems) should be given where possible.
- Key message elements and sequencing should be pre-planned.
- Announcements should be reinforced by message displays where possible.
- Sectoring facilities can help public announcements to be targeted effectively.
- Positive statements and instructions are preferable to negative ones.
- Key items should be repeated (location of problem, required destination, required route, etc).

Chapter 6

Crowd management

279 The safety and enjoyment of people attending a music event will depend largely on the effective management of the crowd. Crowd management, however, is not simply achieved by attempting to control the audience, but by trying to understand their behaviour and the various factors which can affect this. It is necessary to put a complete system into practice rather than attempt to control certain elements of obvious concern, without understanding the underlying issues. Further information on crowd management can be found in HSE's *Managing crowds safely*.

280 In addition to the aspects covered in this chapter, many other factors in the design and planning of the event, discussed elsewhere in this publication, will have a bearing on crowd management, such as:

- design of the venue to allow good entry and exit and to allow for crowd movement within the venue;
- audience capacity;
- provision of adequate facilities for refreshments, sanitary requirements, etc;
- clear, effective means of communication with the audience.

Audience profile and crowd dynamics

281 Two important aspects to be considered in crowd management are:

- audience profile; and
- crowd dynamics.

282 Many factors may introduce the potential for crowd movement and therefore need to be considered at the venue and site-design stage, such as:

- multiple-stage entertainment;
- provision of satellite stages, platforms and stage thrusts;
- sound and video towers;
- sight-line obstructions or restricted views;
- multiple-barrier systems and pens;
- location of facilities;
- the psychological state of the audience;
- special effects.

283 The way in which crowds behave and respond is a combination of physical and other factors. The dynamics of the crowd will depend, in a large part, on the activities of the crowd and this, in turn will be influenced by the character of both the crowd and the groups or artists performing.

284 Matters to be addressed include:

- the character of the artists or groups, eg diving into audience, throwing items into audience and performing in audience arena;
- the audience profile, eg male/female split, age of audience, heavy consumption of alcohol or likelihood of drug consumption, physical behaviour, eg 'slammers';
- likely crowd activities; eg body surfers, slam dancers, moshers, aerialists and stage diving.

285 It is important for stewards to be able to recognise and understand what are 'normal' activities for the audience.

Entry and exit of the audience

286 Before the audience enters the venue, ensure that checks are made of all fire and emergency facilities and that:

- all exits are unlocked;
- escape routes are clear;
- emergency lighting works;
- fire-fighting equipment and alarms are in full working order;
- a PA system for use in emergencies can be heard clearly in all parts of the venue.

If these checks are to be carried out by stewards clear instructions must be given.

Entrances and exits

287 Ensure that entrances and exits are clearly signposted and operate efficiently. Consider the needs of children and people with mobility difficulties. Separate entrances and exits for pedestrian access from entry routes used by emergency services, and concession vehicles. Provide information to the audience about any restricted exits that are not in use while the event is in progress (see chapter on *Venue and site design* for more information on entrances and exits).

Opening time

288 Problems may occur at entry points if large numbers of people seek to gain admission at the same time and if the situation is not properly managed this may result in crushing injuries. It is therefore recommended that:

- entrances are opened some considerable time (eg 1-2 hours) before the event is due to start and the audience is made aware of this by tickets, posters or other means. If significant crowding is likely to occur before that time, consider opening gates before the published time, providing that on-site services are ready;
- admission is staggered by providing early supporting acts or other activities.

289 It is important to appreciate that when entrances are opened early the audience demands on facilities such as waste clearing, sanitary accommodation and catering, will be increased.

Crowd pressure at the entrances

290 This can be reduced by:

- keeping all other activities, including mobile concessions, well clear of entry points;
- arranging for adequate queuing areas away from entrances;
- creating holding areas away from entrances to relieve the pressures on these points;
- ensuring that barriers, fences, gates and turnstiles are suitable and sufficient for the numbers using them;
- locating ticket sales and pick-up points away from the entrance;
- providing a sufficient number of trained and competent stewards;
- arranging for a short-range PA system and megaphones to be made available at entrances to notify people of any delay.

Opening the entrances and arrangements for the front-of-stage area

291 When entrances are first opened at non-seated events, the audience tends to rush towards the front which can cause tripping accidents and injuries. Carefully consider how the area in front of the stage will be managed and stewarded when the entrances are opened. If a standing area is provided in front of the stage, make sure that entrances do not lead directly to this area from stage right or left.

292 One recommended method of easing the initial rush towards the stage and preventing slipping or tripping accidents is to provide a line or lines of stewards across the arena through which the audience can move towards the stage in an orderly manner. This may be supplemented by PA announcements to keep the audience informed about what is happening.

Ticketing

293 Ticketing policies can have a direct effect on the safe management of the audience. Consider the following:

- where a capacity or near-capacity attendance is expected for an event, admission should be by advance ticket only;
- tickets for seats which offer restricted views, or are uncovered, are marked accordingly, and the buyer forewarned;
- tickets for seats with severely restricted views are not sold;
- part of the ticket retained by the audience member after passing through a ticket control point should clearly identify the location of the accommodation for which it has been issued;
- a simplified, understandable ground plan is shown on the reverse side;

- if there is more than one entrance, introduce colour coding of tickets corresponding to different entrances and ensure audience members are proportionally divided between entrances;
- all sections of the venue, all aisles, rows and individual seats, are clearly marked or numbered, as per the ticketing information.

Admission policies

294 As stated above, the admission policies can have a direct effect on the rates of admission and the management of entrance areas and audience accommodation in general. Specific points to be considered include:

Cash sales

295 To ensure a steady flow of audience into the venue when entry is by cash, set the admission price at a round figure. This avoids the need for handling large amounts of small change.

Ticket-only sales

296 The advantage of confining entry to ticket-only is that the rate of admission should be higher than for cash sale. If tickets are sold at the event, wherever possible, provide separate sales outlets. Ensure that these outlets are clearly signposted and positioned so that queues do not conflict with queues for other entry points.

Reserved (or numbered) seat ticket sales

297 Selling tickets for specific numbered seats has the advantage that the seats are more likely to be sold in blocks and the system allows different categories of audience member (eg parent and child) to purchase adjacent seats and enter the venue together. This policy helps to avoid random gaps and ensures that in the key period preceding the start of the event there will be less need for stewards to direct latecomers to the remaining seats, or move members of the audience who have already settled.

Unreserved seat sales

298 Selling unreserved seats has the advantage of being easier to administer. However, people are prone to occupy seats in a random pattern, and, as stated in paragraph 297, it can be hard to fill unoccupied seats in the important period before the start of the event. For this reason, when seats are sold unreserved, a reduction in the number of seats made available for sale may be necessary (in the region of 5-10% of total capacity, according to local circumstances).

No ticket sales on site

299 If all tickets have sold out in advance, or if tickets are not sold on site, every effort should be made to publicise this fact in the media. In addition, place signs advising people of the situation along all approaches to the event, to avoid an unnecessary build-up of crowds outside. This is a preferred method for likely sell-out concerts

Ticket design

300 Ticket design can have a direct effect on the rate of admission. Clear, easy-to-read information will speed the ability of the entry-point steward to process the ticket. Similarly, if anti-counterfeiting features are incorporated (as is recommended), ensure that there are simple procedures in place for the steward to check each ticket's validity.

Admission of young children

301 It may not be appropriate to allow young children, particularly those under the age of five years, to attend certain events because they may be trampled or crushed. If they are not to be allowed in, clearly advertise this fact in advance. Where young children are allowed, consider arrangements for prams and pushchairs, and at large events, dedicated children's areas may be useful. Consider contingency planning for dealing with this element of the audience, such as relocation to a specific area and ensure that you have a procedure in place for stewards to assist with such relocation.

Pass outs

302 Pass outs enable members of the audience to leave the event for a short time and return. Consider this facility for events that will last for more than four hours.

Guest/VIP/restricted areas

303 Separate access points may be needed for particular types of ticket holders, eg guests and VIP's, artists and their entourage, workers, officials and emergency services workers. Consider the location of the gates between these areas and the main arena, to prevent any crowd build-up at such points. Clear identification of people permitted into such areas will assist stewards in controlling admission and in minimising delays in admission, which reduces queuing. Such identification may be by means of special passes or wristbands.

Searching

304 Searching at entrances may be necessary to prevent prohibited items from being brought onto the site. Ensure that searching is only carried out by properly trained and supervised stewards.

Late leavers

305 At the end of the event when most of the audience have left, if practicable, stewards can form a line in front of the stage and slowly walk to the furthermost exit, moving the remaining audience out of the area.

Crowd sway/surges

306 At large events it is sometimes effective to subdivide the audience into pens, which reduces the effects of sway and surge. If this method is used, ensure that there is a system in place to prevent overcrowding.

307 Think carefully about where to position stewards to monitor the audience for distress, crushing, sway, or surges, as they all present a risk to members of the audience. Use of CCTV and/or the provision of raised viewing platforms, especially stage left and stage right, may help to monitor the audience for signs of distress.

308 If people are at risk, you will need to take immediate action, eg by enlisting the assistance of performers and by making an announcement. The performers could be asked to alert you or the safety co-ordinator if they are concerned about a possible serious audience problem. It can then be investigated immediately.

Police involvement

309 If there is to be a police presence in, or at the event, the responsibilities and functions of the police need to be agreed and documented, eg whether particular posts are to be staffed by stewards or by police officers, and who will assume responsibility in particular circumstances. Record the outcome of these discussions in a statement of intent. Remember that a statement of intent is a management statement and not a legal document.

Aids to crowd management

Use of PA systems and video screens

310 It may be helpful to arrange a safety announcement for the audience before the event starts. The announcement could give information about the location of exits, the identification of stewards and procedures for evacuation. The use of video screens to provide entertainment before the event and during changeover periods can also help crowd management. They can be used to inform the audience about safety arrangements, facilities on the site and transport, etc. However, screens may not be visible in all parts of the site and so it may be necessary to plan supplementary means of giving information.

Stewarding

311 The main responsibility of stewards is crowd management. They are also there to assist the police and other emergency services if necessary. Apart from the specialist workers provided for the protection of the performers, the use of separate teams for security and stewarding should not be considered without consultation between all interested parties. The roles of these two groups are closely inter-linked and lack of communication can lead to ineffective crowd management.

Deployment and numbers of stewards

312 The risk assessment will help you to establish the number of stewards necessary to manage the audience safely. When preparing your risk assessment for crowd management, carry out a comprehensive survey to assess the various parts of the site and consider the size and profile of the audience.

313 Basing stewarding numbers on the risk assessment rather than on a precise mathematical formula will allow a full account to be taken of all relevant circumstances, including previous experience. To manage the audience, locate stewards at key points. These include barriers, pit areas, gangways, entrances and exits and the mixer desk and delay towers.

314 An example of some of the matters to be considered for the risk assessment include:

- previous experience of specific behaviour associated with the performers;
- uneven ground, presence of obstacles, etc, within or around site, affecting flow rates;
- length of perimeter fencing;
- type of stage barrier and any secondary barriers;
- provision of seating.

Further information regarding risk assessments for crowd management can be found in the document *Research to develop a methodology for the assessment of risks to crowd safety in public venues.*

Organisation of stewards

315 There has to be an established chain of command. Consider appointing a chief steward to be responsible for the effective management of all stewarding contractors at the event. (This could be a role of the safety co-ordinator.) The arrangements will depend on the nature and size of the event and venue but may include:

- a chief steward;
- a number of senior supervisors, responsible for specific tasks, who report directly to the chief steward; and
- a number of supervisors who report direct to a senior supervisor and who are normally in charge of six to ten stewards.

316 Ensure that stewards receive a written statement of their duties, a checklist (if this is appropriate), and a plan showing key features. Brief stewards before the event, particularly about communicating with supervisors and others in the event of a major incident.

Conduct of stewards

317 All stewards need to be fit to carry out their allocated duties, aged 18 years and over, and while on duty they should concentrate only on their duties and not on the performance. Ensure that stewards understand that they should:

- not leave their place without permission;
- not consume or be under the influence of alcohol or other drugs; and
- remain calm and be courteous towards all members of the audience.

318 All stewards should wear distinctive clothing, such as tabards and be individually identifiable by means of a number which is clearly visible.

Competency of stewards

319 Duties and competencies of stewards include:

- understanding their general responsibilities towards the health and safety of all categories of audience (including those with special needs and children), other stewards, event workers and themselves;
- carrying out pre-event safety checks;
- being familiar with the layout of the site and able to assist the audience by giving information about the available facilities including first aid, toilet, water, welfare and facilities for people with special needs, etc;
- staffing entrances, exits and other strategic points; eg exit doors or gates which are not continuously secured in the open position while the event is in progress;
- controlling or directing the audience who are entering or leaving the event, to help achieve an even flow of people into and from the various parts of the site;
- recognising crowd conditions to ensure the safe dispersal of audience and the prevention of overcrowding;
- assisting in the safe operation of the event by keeping gangways and exits clear at all times and preventing standing on seats and furniture;
- investigating any disturbances or incidents;
- ensuring that combustible refuse does not accumulate;
- responding to emergencies (such as the early stages of a fire), raising the alarm and taking the necessary immediate action;
- being familiar with the arrangements for evacuating the audience, including coded messages and undertaking specific duties in an emergency;
- communicating with the incident control centre in the event of an emergency.

Stewards' training

320 Ensure that all stewards are trained so that they can carry out their duties effectively. The level of training will depend on the type of functions to be performed. Keep a record of the training and instruction provided, including the:

- date of the instruction or exercise;
- duration;
- name of the person giving the instruction;
- name of the person(s) receiving the instruction; and
- nature of the instruction or training.

321 All stewards need to be trained in fire safety matters, emergency evacuation and dealing with incidents such as bomb threats. For those working in the pit area, make sure they are trained so that they are able to lift distressed people out of the audience safely and without risk to themselves. (The *Guide to safety at sports grounds* provides some further information on the training of stewards working in football stadia.)

Stewards' welfare

322 Ensure that stewards are not stationed for long periods near to loudspeakers and make sure they are provided with ear protection in accordance with the Noise at Work Regulations 1989 (see chapter on *Sound: noise and vibration*). Stewards will need adequate rest breaks so ensure that arrangements are in place for them to have rest periods at reasonable intervals.

Chapter 7

Transport management

323 Traffic management proposals need to be planned to ensure safe and convenient site access and to minimise off-site traffic disruption. Set your traffic management proposals out in a transport management plan and agree the plan with the police and local highway authority.

Traffic signs and highway department road closures

324 Identify the need for temporary traffic signs before the event. If temporary traffic signs are needed, prepare and agree detailed traffic signs plans and schedules with the police and local highway authorities before the event. It may be necessary for people living in the area to be consulted over route changes and to be advised of the impact, once agreement has been reached. Consider using a traffic sign contractor for events where the majority of people will be arriving by cars or coaches.

325 Consider the need for temporary traffic regulation orders to provide for road closures, banned turns, lane closures, parking restrictions, temporary speed limits and lay-by closures. For large events and particularly if special traffic management arrangements and temporary traffic regulation orders are required, consultation with the local highway authority is essential. Highway authorities include the highways agency for all trunk roads and motorways, and the local authority for all other roads.

326 Consult the local highway authority as to the best way of carrying out traffic orders and allow sufficient time for any temporary traffic regulation orders to be processed.

Traffic marshalling

327 Only the police or someone under their direction can legally undertake traffic regulation on the public highway. Consultation is therefore essential to secure the appropriate provision of resources. Stewards directing traffic on site should have suitable personnel protective equipment such as high-visibility clothing and weather protection. Stewards should receive traffic marshalling training, eg safe positioning of the marshal and awareness of visibility problems for drivers of reversing vehicles.

328 Make sure that there is suitable and sufficient communication between on-site and off-site traffic marshalling regarding temporary one-way systems, etc. Also, provide adequate numbers of stewards to manage the traffic flows and deal with the parking of vehicles.

Public transport

Trains and underground trains

329 If appropriate, consult with rail authorities about introducing additional trains or enhancing existing services to accommodate the demands of the event and to limit the demand for on-site and off-site parking.

330 It may also be worth investigating the use of combined event/rail package tickets. Consideration, however, needs to be given to the distance between railway stations and the venue and the availability of connecting bus and coach services to and from the event. Advertising may be carried out on trains and at stations before the event, stating any additional service (or lack of) being provided.

331 It is also important to consult the rail authorities concerning the maximum number of people that a station can accommodate at any one time. Most railway stations will have contingency plans, which identify the safe number of people allowed on the platform at any one time. These contingency plans can be used at the event planning meetings between the relevant railway authorities, police, British Transport police and local authority.

332 The British Transport police are responsible for crowd control on railway property. Train-operating companies and Railtrack have responsibility for the queuing of large numbers of people at their stations. Plan how you are going to communicate with the train-operating companies and the police in the event of a major incident to ensure that the stations receive advance information in case the event finishes earlier than planned or emergency evacuation is necessary.

Public transport management

Advice to train-operating companies

333 Train-operating companies need to consider their own planning procedures to ensure that they can safely manage the potential increased 'through put' of passengers associated with the event, eg ensuring suitable entrances and exits, control of passenger numbers on platforms, footbridges and tunnels, crowd flow plans and temporary queuing system and communicating travel information by PA systems.

334 It is important that the transport providers also draw up their own contingency plans for dealing with train delays or incidents on the track and consider the suitability of the rolling stock, provision of first-aid points and first aiders, additional toilets and additional workers.

Coaches/buses

335 Planning the arrival and departure of coaches can greatly reduce congestion at the beginning and end of a large music event. Careful consideration has to be given to the routing of such vehicles. Parking areas and access roads should be provided to reduce as far as possible the need for coaches/buses to reverse, eg creating one-way systems.

336 Coaches need wide and easily accessible entrance and exit points, as well as large turning areas into allocated parking areas. Consider specific arrangements to ensure the free flow of coach routes in consultation with the police, and ensure that this is documented in the transport management plan. Coach parking areas may need to contain toilet facilities.

337 Private bus/coach operators are often prepared to provide special shuttle bus services between local rail and/or bus stations. However, shuttle bus systems may not be appropriate for all events. Congestion caused by a natural mass exodus at the end of an event is likely to prohibit free flow of traffic routes and consequently shuttle buses become unable to operate effectively. Consider the potential for dedicated shuttle bus routes or consult local bus operators about enhancing or extending their established services to serve any proposed event.

Vehicle parking and management

338 Include proposals in the transport management plan for the management of vehicle parking which identify the likely resources required (space necessary, traffic marshals and equipment) and methods to be used for parking management.

339 Make sure that both you and the police can communicate with the vehicle parking management team, so that resources can be directed quickly to deal with any incidents within the car parks or at the various site accesses.

340 For large events, consider the appointment of a traffic management co-ordinator who will liaise with the police, car park management, traffic signs contractor, local highway authority and local authority.

Vehicular access

341 Ensure that the road signs are appropriate and easily visible, the capacities of the parking areas are adequate and the surface is capable of withstanding the anticipated traffic volume. Consider using hardcore, trackway and/or other suitable temporary surfacing which can prevent damage to the ground and prove invaluable in wet ground conditions.

342 Detailed capacity assessments may be needed to ensure access entry capacity is adequate. Queuing on entry into the site can cause blocking of traffic flows leading, potentially, to severe congestion. Exit capacity is less problematic as, if congestion occurs on exit, it is contained within the site and will not adversely affect off-site conditions. However, the risks associated with poor vehicle exit management should not be underestimated. Methods for ensuring the safe exit of vehicles from the site need just as much careful planning. Consider planning alternative routes and accesses. These can be used if main access points or routes become blocked.

343 Consider vehicle access for service vehicles before, during and after the event, eg waste collection vehicles and sanitary servicing vehicles.

Parking

344 Consider separate parking areas for the general audience traffic, vehicles for people with special needs (close to event site), coaches, shuttle buses, guests/VIPs, artistes, emergency service workers and event workers. Overspill-parking facilities either on site or at a convenient location off-site to accommodate the potential for excess visitors may also need to be planned. This may take the form of a vehicular circulation/holding area as a temporary measure.

345 Car and coach parks need to be adequately lit, signposted and labelled with reflective numerals or letters so that vehicles can be easily located at the end of an event or in any other emergency. Ideally, separate coaches from car parks. For large outdoor events, position signs at exit gates leading from the parking area to the venue to assist in identifying where cars have been parked and consider clear signs for exiting vehicles showing route direction.

Emergency access

346 Plan provision for the entry and exit of emergency service vehicles. Ideally these routes should be separate and safeguarded. The routes and access chosen must allow for means of access by the fire brigade to within 50 m of any structure, including fuel storage facilities. The access route will need to bear the weight of fire appliances and avoid manhole covers. These routes should be signposted.

347 Get advice from the fire brigade concerning access route specification and incorporate this into the transport management plan. In this respect, early application for road closures and temporary traffic regulation orders may be necessary. It is also important to identify allocated emergency vehicle rendezvous points in the transport management plan.

Pedestrians

348 Identify safe means of entry and exit for pedestrians, ideally segregated from vehicular access. Where pedestrian access is difficult, consider the provision of alternative means of access, eg shuttle buses to collect pedestrians en route. Consider making specific arrangements for those attending who have a physical disability and may not be able to walk long distances. Avoid entry and exit routes crossing car or coach parks and traffic routes. Where the latter is unavoidable, plan for adequate traffic control measures.

On-site vehicle management and temporary roadways

349 It is important to minimise traffic movement within the site and conflicts between vehicles and pedestrians. Consider moving vehicles into the parking areas as efficiently as possible and having a dedicated access to parking areas with no ticket checks on entry. In some circumstances, ticket checks can be undertaken on pedestrian exits from the parking into the event area. This may, however, not be practicable for camping events.

350 Restrict traffic movement in the event arena to emergency service vehicles and other essential services. Consider speed restrictions on site and plan separate access for production vehicles.

351 Temporary roadways are useful to allow suitable hard-surfaced access for pedestrians and service vehicles. Plan temporary access roads, ideally to provide for two-way emergency access or one way with passing places and working space as appropriate. All on-site vehicles must display adequate lighting at night-time and remember to keep pedestrian and vehicle conflict points to a minimum. Plan how the vehicles that will be delivering equipment and provisions are to enter and exit the site safely during the 'build up' (and breakdown) of the event.

352 Where vehicle routes change from those arranged at planning stage, due to heavy rain or some other unforeseen circumstance, make sure that arrangements are in place for reinforcing the alternative route. Safe vehicle recovery from soft ground should be planned.

Lift trucks and other vehicles

Lift trucks

353 No one should be permitted to operate a lift truck unless they have been selected, trained and authorised to do so. The HSE publication *Rider-operated lift trucks: Operator training Approved Code of Practice and guidance* provides practical guidance in relation to the necessary training.

354 Trained operators will have a certificate from an accredited organisation indicating the type of lift truck for which they have received training. A certificate to drive one lift truck does not qualify an operator to drive other types of lift truck. Do not allow workers to operate lift trucks without checking that they are fully trained for the type of truck they are to use.

355 If lift trucks are hired, check that the equipment is delivered in a safe working condition. They should be marked with their safe working load to comply with the Lifting Operations and Lifting Equipment Regulations 1998 (LOLER) (see *Safe use of lifting equipment* and paragraph 382). These Regulations also require safe working practices when using trucks. Ensure they come with a current report of thorough examination which covers adequately the period they will be used on the site.

356 The Provision and Use of Work Equipment Regulations 1998 (PUWER) cover the use and maintenance of lift trucks (see *Safe use of work equipment)*. Both LOLER and PUWER require periodic inspections of the vehicle and its lifting equipment.

Other vehicles used on site

357 As well as lift trucks, there is likely to be the need for other types of vehicles to operate on site such as:

- other specialist lifting vehicles, eg scissor lifts;
- vehicles used to deliver equipment around the site or venue, eg golf buggies and electric carts; and
- other vehicles, eg tractors, trailers and waste-collection vehicles.

358 If these vehicles are being used for work activities, PUWER and in some cases LOLER may apply. The use of all vehicles on site should therefore be carefully planned and monitored to ensure that accidents do not result from the incorrect use of the vehicle or that pedestrians are not injured as a result of their use.

Chapter 8

Structures

359 Many events require the provision of temporary demountable structures, eg grandstands, stages, marquees. Managing the hazards connected with these structures is just as important as managing other hazards. This can only be achieved if all those responsible for these structures undertake their duties conscientiously.

360 The failure of any temporary demountable structure, no matter how small, in a crowded, confined space could have devastating effects. It is therefore essential to design and erect structures to suit the specific intended purpose, and to recognise that the key to the safety of these structures is largely in the:

- choice of appropriate design and materials;
- correct siting or positioning;
- proper planning and control of work practices; and
- careful inspection of the finished product.

361 This chapter gives guidance on providing safe temporary demountable structures. It starts with the preliminary decisions that need to be made - choosing the site and the supplier - and continues to give general guidance on:

- the safety requirements for temporary demountable structures;
- the documentation required to ensure that the essential safety requirements are provided;
- advice on post-erection management of temporary demountable structures.

362 While it is intended that this chapter will complement the Institution of Structural Engineers

document *Temporary demountable structures: Guidance on design, procurement and use*, it is not meant to be used instead of the Institution's document and other relevant design standards or vice versa.

Scope

363 The kinds of structure usually found at music events include stages, sets, barriers, fencing, tents and marquees, seating, lighting and special effect towers, platforms and masts, video screens, TV platforms and crane jibs, dance platforms, loudspeaker stacks, signage and advertising hoardings. This chapter applies to temporary demountable structures erected indoors and outdoors. Temporary demountable structures erected outdoors will need to meet all the requirements of indoor structures plus the additional factors created by the effects of the weather.

How the law applies

364 In general, the erection and dismantling of temporary stages, grandstands and other temporary platform arrangements used by the entertainment industry are not construction operations. The specific requirements of the various regulations made under the Health and Safety at Work etc Act (HSW Act), specifically for construction, will not apply to this kind of work. Similarly, the Construction (Design and Management) Regulations 1994 do not apply to these kinds of structure.

365 However, other law will apply. The HSW Act and the Management of Health and Safety at Work Regulations 1992 (Management Regulations) apply. The responsibility for the enforcement of the HSW Act and associated regulations is therefore the responsibility of the health and safety enforcing authority for the whole event. In most cases this will be the local authority health and safety inspectors unless the local authority are themselves organising the event in which case it will be the HSE.

366 Risks should be assessed, hazards eliminated or reduced and safe systems of work developed. The various regulations relating to construction work will assist in identifying the types of precautions which need to be taken to control the risks associated with this type of work. In Scotland, the Civic Government (Scotland) Act 1982 controls the use of temporary structures and approval must be sought from the local authority.

Preliminaries

Choosing the location

367 The following factors may influence the choice of location for temporary structures.

- Is the site adequately drained? If the site is liable to flooding, this could cause either the load bearing capacity of the ground to be reduced or wash away the ground under the supports. Take measures to control the effects of this.
- Is the site flat or can it be made flat? Where there is a gradient or the ground is uneven, the structure needs to be capable of being modified to deal with such variations.
- Are there overhead power cables, and if so are they sufficiently clear of the upper part of the structure (or cranes which may be employed in the assembly of temporary demountable structures)?
- Does the proximity of surrounding buildings, structures and vegetation create risks in relation to the possible spread of fire?

368 Obtain information about the load bearing capacity of the ground or floor. For outdoor events ensure that the ground load bearing capacity is capable of supporting the imposed loadings in all weather conditions. Indoor venues may have gaps or basements under the floor surface. Very high point-loads may be created by the use of cranes or lift trucks to install sections of structures or equipment and specialist advice is essential.

Choosing the supplier

369 Choose a competent supplier for all temporary demountable structures to be erected and used on site. A competent supplier will be able to demonstrate at least the following:

- a knowledge and understanding of the work involved;
- that they can manage/eliminate the risks involved in constructing these types of structure;
- that they employ a suitably trained workforce.

370 It is important to note that the design of temporary demountable structures is outside mainstream civil and structural engineering. Therefore, the design of temporary structures should only be carried out by suitably competent people. A suitable designer must be able to demonstrate:

- a full understanding of the loads that these types of structure may be subjected to;
- a full understanding of the properties of the materials normally used for these structures;
- a knowledge of the skills of the people normally employed to erect these structures;
- a full understanding of the proprietary structural elements used in these structures.

Essential requirements

Design

371 All temporary structures must possess adequate strength and stability, in service and during construction. The means of achieving these are covered, in some detail, in various British Standards and other guidance. Further information can be found in the document *Temporary demountable structures: Guidance on design, procurement and use.*

372 The design of a temporary structure should provide protection against falls for:

- performers - consider the need for handrails at an appropriate height for all stage areas, platforms and access ways;

- workers;
- the audience.

373 In addition, the surface of any ramp or tread, particularly those which could become wet, should be covered with a slip-resistant material.

Erection

374 To prevent the incorrect erection and subsequent use of temporary structures, attention should be paid to the following.

- The assembly of temporary structures should be carried out in accordance with calculations, plans and specifications drawn up by a competent designer.
- Apparent similarities between proprietary systems used for temporary structures may only be cosmetic and it is important not to mix products from different manufacturers unless the potential implications have been fully considered.
- Erection should take place in a way that ensures stability at all times.
- When practicable, temporary structures should be erected either from the ground or from a suitable stable platform.
- Many temporary demountable structures cannot be built except by climbing the framework as it is assembled and this should be addressed in the risk assessment and safety method statement.
- Equipment should be checked to ensure that it is fit for its purpose and fully meets any specification which has been laid down, eg steel items with cracked welds, bent or buckled members, or with large amounts of corrosion should be rejected.
- All components should be examined during assembly (and dismantling) for signs of wear, deformation or other damage, and replaced where necessary.
- Correct alignment of components is important - they should not be bent, distorted or otherwise altered to force them to fit.
- Particular attention should be given to fastenings and connections. It is essential to provide suitable covering for bolts and fittings which project into or adjoin audience areas.
- Earthing.
- Where guying is used, care should be taken to ensure that the guys and their anchors do not cause an obstruction. All stakes or anchors should be located or covered so that they do not create a tripping hazard.

Protecting erectors against falling

375 Virtually every temporary demountable structure is free-standing without the benefit of support from existing buildings or similar. Therefore it is very difficult to provide effective fall restraint systems for the workers assembling or dismantling the top components. In common with the construction scaffold industry, maximum protection is afforded by the selection of competent workers who have demonstrated their aptitude for the task and are subject to ongoing assessment and training as appropriate.

376 Where personal protective equipment is assessed to be the most effective means of controlling the risk of injury, employers must issue this to workers, advise on its use and ensure that it meets the requirements of the Personal Protective Equipment at Work Regulations 1992. (See HSE's *Working at heights in the broadcasting and entertainment industries,* Entertainment Sheet No 6.)

Protection from falling objects

377 While structures are being erected, try not to lift materials over the heads of people working or passing below. Create 'no go' areas below working areas from which other workers are excluded.

Limiting the loads carried by people

378 The Manual Handling Operations Regulations 1992 have replaced all existing legislation on the lifting and carrying of loads. They set out new requirements for safe handling of loads where there is a risk of injury and cover all hazardous manual handling operations.

379 The main duties of employers are to:

- so far as is reasonably practical, avoid the need for manual handling of loads involving a risk of injury;
- assess the risk of injury in those operations that cannot be avoided;
- reduce the risk of injury to the lowest level reasonably practicable using the assessment as a basis for action.

380 The assessment must take into account a number of factors including the load, the task, the working environment and individual capability.

Use of lifting and rigging equipment

381 The Lifting Operations and Lifting Equipment Regulations 1998 (LOLER) apply in all the premises and work situations subject to the HSW Act and build on the requirements of the Provision and Use of Work Equipment Regulations 1998 (PUWER).

382 Any organisation using lifting equipment has a duty under LOLER to provide physical evidence (eg a copy of the last report of thorough examination) to health and safety inspectors to demonstrate that the last inspection has been carried out. People hiring lifting equipment should make sure that it is accompanied by the necessary documentation. After positioning rigging and similar equipment, the user should ensure that a competent person inspects the lifting equipment before it is put into use to make sure it is safe to operate. The user then has the duty to manage the subsequent lifting operations in a safe manner.

383 The selection of suitable work equipment for particular tasks and processes makes it possible to reduce or remove many risks to the health and safety of people at the workplace. This applies both to the normal use of the equipment as well as to other operations such as maintenance. The risk assessment will help to select work equipment and assess its suitability for particular tasks.

384 Everyone involved in erecting and dismantling temporary demountable structures must be appropriately trained. Training is now commercially available in safe techniques for high level rigging, and those working at high level must have undergone training and assessment.

Dismantling

385 Dismantling of temporary demountable structures is subject to the same risks as the assembly operation. Therefore, it should be carried out methodically by people who are appropriately trained and strictly in accordance with the design documentation. Items or components should be handed or lowered down, never dropped or 'bombed'.

Essential documentation

Design concept and statement

386 All proper designs will have calculations to determine the balance of loading and scale of forces acting on the structure. Therefore, the designer should be able to provide:

- a statement as to what the structure is designed to do (the concept);
- a list of items or connections that require particular checking each time the structure is erected;
- particularly for outdoor structures, details of the methods of transferring all horizontal forces, eg wind, back to the ground (without which the structure will not be stable).

387 The physical checking of temporary structures becomes much more effective and simple if the designer's statement is available to the local authority.

Construction drawings

388 Construction drawings will normally be required for all but the simplest temporary structures. These should be accompanied by full calculations, design loads and any relevant test results. These documents should normally be sent to the local authority at least 14 days before the event. It should be recognised that supplementary details, eg loads from lighting and sound suppliers, may not be available until nearer to the event.

Risk assessment

389 A risk assessment should be carried out by the contractor to cover the erection and design of the temporary demountable structure. Remember that the effort and resources applied to health and safety issues should be proportional to risks associated with the project and the difficulty of managing those risks. It may be necessary to carry out another risk assessment to consider the hazards that the temporary demountable structure may create by being in the venue.

Safety method statement

390 A safety method statement should be drawn up for the erection and dismantling of any structure. This should be submitted with the initial plans and calculations to the local authority. The method statement should be specific to the structure.

Completion certificate

391 Monitor all activities at the venue relating to the erection and construction of temporary demountable structures to ensure that they are erected to the detailed specification and that safety method statements and safe working practices are followed.

392 Ensure that all structures are checked by a competent person after they have been erected and before they are used, to make sure that they conform with the drawings and specified details. If this check is carried out by someone employed by the contractor erecting the structure, verify that the checks have been carried out effectively and have been recorded. At this stage, a completion certificate may be issued. A completion certificate is a statement that the work has been carried out in accordance with the designer's specification.

393 If self-certification is used, it is unlikely that the local authority will carry out any inspections of the temporary demountable structure. It is therefore critical to ensure that each contractor

certifies their structure/s as complete and that this documentation is passed to the local authority.

The temporary demountable structure must be certified as complete only by people who are competent to do so.

Managing the completed structure

Before admitting the audience

394 Temporary demountable structures should comply fully with the design documentation, before the audience is admitted to the site. If modifications to the structure are necessary, liaise fully with the local authority and the designer.

Monitoring after erection

395 A competent person should monitor a structure which is susceptible to the effects of the weather and/or misuse (by overloading the roof structure for instance) at all times. In practice this means that a representative of the supplier, or other suitably qualified person, should be on-site at all times while the temporary demountable structure is in use, either by workers or during a performance. Regularly check the ground after the structure has been erected to confirm that no deterioration in its load bearing capacity, such as excessive settlement, has occurred.

Protection against falls: people and objects

People at work

396 If work is to take place on a completed structure at height, a safe access system will be needed to ensure that maintenance and adjustments can take place. Guard rails for platforms should normally be provided where a drop exceeds 2 m. Where an access platform is not practicable, an alternative is to provide safety nets or a safety harness which can protect workers from falling from working areas.

Objects

397 Platforms over 1.8 m in height should have either a clear space around them or a method of preventing objects falling onto people. Where items are being passed up a structure, eg by means of a line or a lift, no one should be allowed in the area immediately below or next to the loads.

The audience

398 Protection against falls, provided for the audience, must not, in any circumstances, be removed, altered or tampered with in any way.

Providing adequate lighting

399 Lighting should be sufficient to enable people to move safely about on the temporary demountable structure. Avoid dazzling lights and distracting glare. Stairs should be well lit in a way which ensures that shadows are not cast over the main part of the treads. Where necessary, provide local lighting to supplement the general level of lighting available, eg at locations of high-risk, such as where there are unavoidable changes in level.

Note: While workers can be expected to take care, a performer leaving a brightly lit performance area may be unable to see or recognise an exit route, especially if the light level is significantly lower than on stage.

400 Lights and their fittings should be positioned so that they do not themselves form a hazard. Lights should not be allowed to become obscured. Sufficient emergency lighting should be provided in case of partial or complete failure of the normal lighting.

Marking obstructions, edges, etc

401 Fall protection for the edge of the performance area facing the audience is not normally provided but the edge should be clearly marked. Other physical obstructions, unprotected edges, edges by gaps and stair nosings should all be marked with white or luminous tape. Any such markings should be a minimum of 25 mm wide to be visible, but 50 mm is preferable.

Protecting the temporary structure

402 The provision of a fence with stewards, or boarding or cladding to a minimum height of 2.4 m, can stop people gaining access to restricted areas. To prevent tampering it is advisable to construct structures in such a way that sections cannot be removed without special tools.

Altering structures

403 Components in temporary structures must not be removed without first consulting the designer. If cladding is added to structures, they may become more vulnerable to wind and may also allow other forces to be transmitted to the temporary structure. Therefore, never add banners or other types of hoarding to a temporary structure without first consulting the designer.

Work near temporary structures

404 Any ancillary operations which are carried out close to temporary structures should not affect the stability of that structure. Where trenches are to be dug these should be placed at sufficient distance from a temporary structure so as not to undermine or adversely affect stability.

Managing the loads

405 Loads on temporary demountable structures can be applied in various ways. It is important to ensure that they do not exceed the design loads. Therefore, adequate measures must be taken to prevent overload by:

- people - due to overcrowding any part of a temporary demountable structure;
- unauthorised additions - eg banners, hoardings, projection screens, scrims, scenic facades, etc. They should never be added to temporary demountable structures without the prior consent of the designer;
- equipment loads - eg lighting, special effects, sound systems, video and TV screens. These can be significant, therefore it is important that the final installation does not exceed levels permitted by the design documentation.

Chapter 9

Barriers

406 Barriers at music events serve several different purposes. They can provide physical security, as in the case of a high perimeter fence at an outdoor concert, or be used to prevent the audience climbing on top of mixer towers, etc. They may also be used to relieve and prevent the build-up of audience pressures, eg a properly constructed front-of-stage barrier enables those suffering physical distress to be reached and helped more easily.

407 Barriers will always be subject to loading and should therefore be designed to withstand right angle and parallel loads in line with the probable pressures. Account should be taken of the nature of the loading, eg surging. Detailed technical requirements for the various types of barrier referred to in this chapter are given in the book *Temporary demountable structures: Guidance on design, procurement and use.*

Front-of-stage barrier

408 Assess whether such a barrier is needed and what form it should take. If audience pressure is expected a front-of-stage barrier will be necessary. Factors to be taken into account include audience density, the likely behaviour and size of the audience and the nature of the venue. For most concerts, some form of front-of-stage barrier will be required.

409 Audience pressure is normally greatest at the front-of-stage barrier. If the audience surges, dynamic loads may be considerable, but such pressure is momentary and to date has not been identified as the cause of serious injury. First-aid treatment from audience pressure will normally be for fainting and exhaustion often due, in part, to other factors (heat, alcohol, hysteria, etc). However, there is a risk that the audience may 'collapse' due to surging or heaving motions near the front of the stage, resulting in people falling to the ground and being trampled and perhaps asphyxiated. A suitably designed and constructed barrier arrangement can help to reduce the risk of collapse.

The pit

410 The area between the stage and the front-of-stage barrier (the 'pit') should be designed to assist the work of stewards, first aiders and paramedics. An important role of stewards is to extract members of the audience who are in distress. The pit should have a non-slip unobstructed working area behind the barrier which is large enough to allow those in the pit to lift members of the audience into it. Some form of elevated platform inside the barrier can help with the lifting of people and enable stewards to oversee the audience and identify anyone in distress. Entrances or exits from the pit should be unobstructed to allow stretcher-bearers clear access to a medical or first-aid point away from the pit area. It is also helpful if pit exits are at least 1.1 m wide.

411 The pit area should be kept clear of anyone other than stewards and first-aid staff. Any arrangements for TV film crew or photographers to work in the pit area should be planned to ensure that their activity will not interfere with the work of stewards or first-aid staff.

412 A concert held 'in the round' with a standing audience requires special arrangements for a pit area. The provision of an unobstructed escape corridor enables members of the audience taken over the barrier to be led away from the pit. However, care needs to be taken to avoid creating a point where people can be trapped between the escape corridor and the barrier. Plan a method to enable people to return into the arena after having been lifted into the pit.

Front-of-stage barrier construction

413 Modern barrier systems are 'A' framed and rely on a tread plate at the front to maintain their stability. They are normally free standing but if used outdoors they may be fixed to the stage structure with couplers. Fixing by couplers is only appropriate if the stage is designed to resist the imposed lateral load.

414 All barriers should be designed to meet the necessary loadings as described in the publication *Temporary demountable structures: Guidance on design, procurement and use.* Checks should be made by a competent person to ensure that, when erected, the barrier meets the design criteria.

415 To prevent injuries from barriers the following matters should be considered.

- Are barriers smooth with no rough edges or trapping points particularly for feet or hands when under load?

- Do they need to be padded?
- Is there likely to be any movement or settlement of the barrier when a load is imposed and could any such movement cause any injury especially to feet or hands?
- To ease the lifting of members of the audience, has the barrier a smooth, curved timber, or steel top?
- Have steps been taken to ensure that there are no sharp or protruding objects from the barrier, eg bolts?
- Do barriers which have a tread plate or floor panel, have a ramped approach or any similar arrangement to reduce the risk of tripping?

416 There should be a reasonable distance between the front-of-stage barrier and the edge of the stage. In no circumstances should it be less than 1 m and should often be considerably more for outdoor events.

Shape of the front-of-stage barrier

417 If a venue has restricted space, a straight barrier is suitable. However, for large concerts, particularly those outdoors, a convex barrier extending into the audience may be preferable. In such circumstances, the barrier should consist of short, straight sections installed at angles to each other to form a curve across the main performance area, extending to the ends of the side stages. It should be erected in conjunction with escapes to right and left of the stage. Concave stage barriers should not be used as people can be trapped between a curved barrier and a straight barrier and are unable to move away. However, at certain events a 'finger' barrier may be appropriate (see paragraph 423 on finger barriers).

418 A curved barrier can provide the following additional safety benefits. It :

- dissipates audience surges away from the centre of the stage;
- assists means of escape;
- provides a wider front row sight line;
- improves performer safety by placing a greater distance between the stage and the barrier, therefore making it difficult for members of the audience to reach the stage; and
- can provide a wider area for stewards and first aiders to operate within the pit.

Barrier around thrusts

419 A thrust is a section of the stage, which projects from the main body of the stage towards the audience. Where thrusts extend into the audience, a barrier should be provided which complies with the design criteria and loading factors for a front-of-stage barrier. It is advisable to construct a thrust in such a way that it does not create poor sight lines. Care should be taken to ensure that such stage designs do not result in concave trapping points from which audience members cannot escape.

420 With less conventional venue layouts that have in-the-round stages, B stages and other satellite performance spaces, it is important to design the barrier systems to avoid penning people in and creating trapping points.

Side-stage barrier or fences

421 The construction of a high side-stage fence to form a sight-line obstruction will ensure that important exits to the right and left of the stage are kept clear and are available for use in an emergency. Such a fence should always be provided for standing audiences. The design and loading criteria for such a fence can be found in the book *Temporary demountable structures: Guidance on design, procurement and use.*

Additional barrier arrangements

422 At large outdoor events, it may be possible to have an additional barrier arrangement to reduce the likelihood of crowd collapse. This could take the form of a finger barrier, extending into the audience or a multiple-barrier arrangement.

Finger barriers

423 If a finger barrier is used, careful design is needed to avoid the creation of trapping points. The barrier should be able to withstand the same crowd loading as the front-of-stage barrier and should have an area which meets the recommendations set out in paragraph 416, and which enables stewards and first aiders to have access to the audience along its length.

Multiple-barrier arrangements

424 For large events, it may be possible to use a multiple-barrier system (ie double or triple barriers in front of the stage). If it is proposed to use such a system, escape arrangements will need to be agreed with the local authority and fire authority. Multiple-barrier systems are not suitable for all venues; for instance, controlled side escapes may be difficult to incorporate in some venues. Penning of audiences in flat, open areas by means other than the arrangements described in the following paragraphs could also create difficulties in evacuation and is considered unsafe.

425 Where double or triple-barrier arrangements are used, the barriers should form a convex curve into the audience with escapes from both ends. The provision of a corridor or area behind each curved barrier will give stewards and first aiders adequate access to the audience along the length of the barriers. The barriers used to achieve this should meet the required minimum loading.

426 With a very enthusiastic audience, it is likely that many of the problems normally encountered at the front-of-stage barrier will be experienced at the barrier furthest from the stage. It is therefore essential that adequate numbers of first aiders and stewards be provided. However, because of the wider sight-line potential (75% in some cases) and the increased distances from the stage, the incidence of audience surge and crushing may be reduced.

Chapter 10

Electrical installations and lighting

427 Electricity can cause death or serious injury to performers, workers or members of the public if the installation is faulty or not properly managed. This chapter gives some general guidance limited to low-voltage installations. Very large loads and high-voltage systems are not covered in this chapter since they require special considerations. In many circumstances the electrical supply may be of a temporary nature, but this does not mean that it can be sub-standard or of an inferior quality to a permanent installation. Only a competent electrician should carry out electrical work.

428 All electrical installations and equipment must comply with the general requirements of the Electricity at Work Regulations 1989. More specific guidance is provided by:

- BS 7909: 1998 *Code of Practice for temporary distribution systems for ac electrical supplies for entertainment lighting, technical services, and related purposes;*
- BS 7430: 1991 *Code of Practice for earthing;*
- BS 7671: 1992 *Requirements for electrical installations* (also known as the IEE Wiring Regulations). This is the most widely used UK standard for fixed electrical installations. Referred to in BS 7909;
- HSE's guidance note GS50 *Electrical safety at places of entertainment* - for smaller venues;
- HSE's booklet HSR25 *Memorandum of guidance on the Electricity at Work Regulations 1989;*
- HSE booklet INDG247 *Electrical safety for entertainers.*

429 The following paragraph 430 provides a general overview of some of the matters to be considered by an event organiser when planning electrical installations within the overall venue design. Electrical contractors working on site should however refer to the specific guidance mentioned above and not rely on the general overview given in this chapter.

Planning

430 Factors to consider when planning the electrical installation include:

- the location of any existing overhead power lines or buried cables;
- the total power requirements for the site;
- access to a network power supply;
- the use of generators;
- earthing;
- positioning of temporary overhead or underground cables;
- the location of the stages;
- the main isolators controlling the electrical supplies to the stage lighting, sound, special effects, emergency lighting, and lifting equipment;
- the location of mixer positions, etc;
- special power supplies for some equipment, eg equipment from the USA which operates on 110 V, 60 Hz;
- power supplies required for hoists, portable tools, etc;
- the electrical requirements for emergency lighting and exit signs;
- power supplies for catering equipment, first-aid points, incident control room, CCTV cameras, etc;
- power supplies for heating or air conditioning.

Installation

431 Where possible locate the main electrical intake and/or generator enclosure where it is accessible for normal operations or emergencies, but segregated from public areas of the venue. Display danger warning signs around the intake or enclosure. The signs should comply with the Health and Safety (Safety Signs and Signals) Regulations 1996, see *Safety signs and signals*.

432 All electrical equipment which could be exposed to the weather, eg consumer units, distribution boards, etc, should be protected by means of suitable and sufficient covers, enclosures or shelters. As far as practicable, all electrical equipment should be located so that it cannot be touched by members of the public or unauthorised workers.

433 On completion of the electrical installation it should be inspected and tested according to the procedures laid down in BS 7909: 1998.

Cabling

434 Temporary overhead cables, whether they are carrying mains voltage, communication, or television signals, should be securely fixed or supported by a catenary wire. The catenary wire and cables should be placed out of reach of members of the public. The catenary wire should be bonded to the earthing system of the cable supported. The cables should be suspended from the catenary wire by means of suitable hangers spaced at regular intervals to provide adequate support to the cable. Advisory notices warning of the location of the overhead cables and the voltage being carried should be clearly displayed.

435 Wherever possible, segregate traffic and cables routes. If this is not possible, a cable height of not less than 5.8 m is advisable to ensure that high-sided vehicles can pass beneath the overhead cables. Fences should be used to segregate roadways from overhead cables running parallel to the roadway to prevent inadvertent contact.

436 If it is necessary to run cables under ground, BS 7671: 1992 gives extensive details, taking into consideration space factors. Cables should be placed far enough underground to protect against:

- crushing by vehicles;
- damage by machinery, equipment or tools;
- other mechanical damage, (eg members of the public).

437 If cables have to be run on the surface they should be protected against sharp edges or crushing by heavy loads, eg by covering with ramps or rubber mats. Ramps should be conspicuously marked to avoid tripping hazards.

Electricity utility cables

438 Overhead or underground electricity supply cables belonging to an electricity supply company may cross the site, or its access roads. If this is so, precautions must be taken to avoid danger from these cables.

439 HSE publishes two documents developed in conjunction with the electricity supply industry, which provide guidance regarding danger from overhead lines, and underground cables. These are:

- *Avoidance of danger from overhead electricity lines;*
- *Avoidance of danger from underground services.*

Access to electrical equipment

440 A clear working space is required to allow access to:

- control switches and equipment;
- amplification equipment;
- special effects equipment;
- follow spots;
- dimmers;
- high-voltage discharge lighting, eg neon.

441 The main control equipment and items specified above should be clearly identified, and their locations marked on a plan to be located in the incident control room.

442 Protect switchgear to prevent access by unauthorised people. Where switchgear is installed in a locked enclosure, specific keyholders should be given responsibility for operating the equipment to comply safely with any request made by the emergency services.

Generators

443 If generators are to be used, consider their location and accessibility for refuelling purposes. Allow for the storage of the fuel, and accessibility for further fuel deliveries. The generator and its fuel should not be accessible to members of the public or other unauthorised people and may need to be fenced. If the venue is located close to a residential area consider the noise-nuisance factor. If this is excessive, silenced generators may be a suitable means of reducing it.

444 BS 7430: 1991 gives guidance on the earthing of mobile generators for outdoor events.

Electricity to the stage area and effects lighting

445 The electricity supply to the stage should be controlled by a switch or switches and installed in a position accessible at all times to authorised people in the stage area.

446 Where possible, provide sufficient fixed socket outlets within the stage area to avoid the use of flexible extension leads and multi-socket outlets. Fixed socket outlets can be either permanent or on properly mounted temporary distribution boards. It is also advisable for equipment to be located within 2 m of a fixed socket outlet to avoid the need for long trailing leads.

447 Any lantern or other suspended lighting equipment should have a suitable safety chain or safety wire fitted. The weight of any flown lighting equipment should not exceed the safe working load of the securing points. No flown or suspended equipment, including lighting bars and amplification equipment, should rely solely on one suspension cable, clamp or bolt. Each means of suspension should be secured to independent fixing points on the flown equipment and the structure.

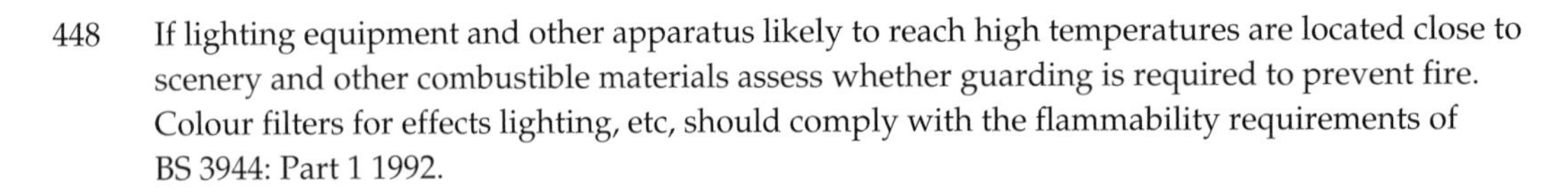

448 If lighting equipment and other apparatus likely to reach high temperatures are located close to scenery and other combustible materials assess whether guarding is required to prevent fire. Colour filters for effects lighting, etc, should comply with the flammability requirements of BS 3944: Part 1 1992.

Normal lighting circuits

449 All parts of an outdoor venue should, unless not intended to be used in the absence of adequate daylight, be provided with suitable levels of artificial light. Consider the lighting of the first-aid post, information area/marquees, pedestrian access to car parks, car park areas, toilets, and access routes to public highways.

Emergency lighting circuits

450 In addition to the normal lighting arrangement, emergency lighting should be provided as determined by the risk assessment and fire-risk assessment. These assessments should cover all possible hazards associated with the venue, eg pits, holes, trenches, ditches, etc. Also consider the provision of emergency lighting within generator enclosures, the main electrical in-take, or main area of isolation.

451 The emergency lighting supply should come from a source of electricity independent of the normal lighting. The emergency lighting should be of a maintained type (continuously lit), which includes

the exit signs located around the venue for directional purposes, and located above the final exit doors. The emergency-lighting arrangements should be in accordance with BS 5266: Part 1 1988; Part 2 1998; Part 3 1981. All exit signs throughout the venue should be in accordance with BS 5499: Part 1 1990 (amd 1995); Part 2 1986 (amd 1995); Part 3 1990. Further information on exit signs can be obtained in the HSE document *Safety signs and signals*.

452 Any source of supply used for providing emergency lighting should be capable of maintaining the full light load as determined by the event risk assessment and the major incident plans prepared for the event, in case of a mains failure. It is important to keep any battery used for this purpose in a fully charged condition whenever the venue is in use.

Management of lighting circuits

453 The normal and emergency lighting systems should be installed so that a fault or accident arising to one system cannot jeopardise the other. Suitable provision should be made to enable repairs to be undertaken if part of these lighting systems fail. Both the normal lighting circuits and emergency lighting circuits, including generators, should be protected from acts of vandalism.

454 Dimming equipment should be located in an approved location, and under the continuous supervision of a competent person when the venue is open to members of the public.

Lighting levels for means of escape

455 All parts of the venue to which people have access should be provided with normal and emergency lighting, capable of giving sufficient light for people to leave safely as determined by the risk assessment. Consider providing additional emergency lighting, operating in a maintained mode to the gangways passing through temporary seating structures. For stairways, gangways/corridors, exit doorways, gates, etc, the average lighting level should be 20 lux and the minimum should be 5 lux.

Portable electrical appliances

456 Portable electrical equipment is defined as equipment which is not part of a fixed installation but is, or intended to be, connected to a fixed installation or a generator by means of a flexible cable and either a plug and socket or a spur box, or similar means. The particular legal requirements relating to the use and maintenance of electrical equipment are contained in the Electricity at Work Regulations 1989. Ensure that any person that may bring portable electrical equipment onto the site can demonstrate that the electrical equipment is maintained correctly and the equipment has been subjected to routine inspection and testing. Further information can be found in HSG107 *Maintaining portable and transportable electrical equipment*.

Chapter 11

Food, drink and water

Catering operations

457 Ensure that the delivery, storage, preparation and sale of food complies with the relevant food safety legislation and where appropriate consideration is given to the advice contained in the relevant industry guides and codes of practice. This will include mobile catering units, catering stalls and marquees, crew catering outlets, hospitality catering, bars and ice cream vendors, etc.

458 Ensure that food businesses carry out their work in a safe and hygienic way. Examine documentary evidence from each caterer regarding:

- the identification and control of potential food hazards by all catering operations;
- the identification and control of potential health and safety hazards by all catering operations;
- provision of appropriate fire extinguishers;
- proper training of all food handlers;
- the suitability of all premises used for the production or sale of food;
- the suitability of the equipment being used;
- transporting food safely and separate from any potential source of contamination;
- storing and disposing of food waste (solid and liquid) properly;
- the maintenance of high standards of the personal hygiene of food handlers;
- the proper storing, handling and preparation of food;
- the provision of a drinking water supply (see paragraph 472);
- insurance of all food businesses including public, product and employers liabilities;

- the possession of electrical and gas installation compliance certificates by all food businesses;
- the possession of a properly equipped first-aid box by each operating unit.

459 Contact the local authority environmental health officers (EHOs) for advice on food safety and hygiene. EHOs may wish to carry out an inspection of the catering facilities provided at the event. They may also require you to provide them with a list of caterers who will be attending the event.

460 Additional requirements may be necessary in certain types of catering operations, eg barbecues and spit roasting. Such operations may present an increased risk of fire, contamination or food poisoning. Carry out a suitable risk assessment, taking into account the particular factors of the operation.

Positioning

461 Your site plan of the event will need to include a detailed layout of all catering operations (see chapter on *Venue and site design*), bearing in mind the need to:

- prevent any obstruction that may affect the health and safety of people attending or working at the event;
- prevent, as far as is possible, access to the rear of catering operations by the audience;
- allow entry and exit for emergency vehicles;
- take into account suitable spacing between individual operations;
- provide readily accessible and preferably lockable facilities for the storage and disposal of solid and liquid waste;
- allow for the efficient removal of refuse (see chapter on *Waste management*);
- position catering operations within close proximity to a supply of drinking water, foul drainage and within a safe minimum distance from any source of possible contamination, ie fuel, waste or refuse storage;
- consider manual-handling issues involved in the disposal of water, the delivery of supplies, etc;
- provide separate toilet facilities for the exclusive use of food-handlers, with hot and cold hand-washing facilities;
- provide suitable facilities for parking and access of support vehicles;
- position mobile sleeping accommodation away from the catering operations.

LPG

462 Liquefied petroleum gas (LPG) is the main source of fuel for outside catering operations. It does present a substantial fire/explosion risk, therefore ensure that:

- all operators using LPG can demonstrate a basic understanding of its safe use, its characteristics and emergency procedures;
- storage at each catering operation does not exceed that which is required for a 24-hour period or a maximum of 200 kg, whichever is the least;
- all LPG is handled and stored in accordance with the current regulations and codes of practice;
- all supplies of LPG whether in compounds or within catering operations are secure from interference by the audience.

Electrical installations

463 Electrical power to catering operations should, wherever possible, be provided by the site electrical supply (see chapter on *Electrical installations and lighting*). If portable generators are used, preference should be given to LPG or diesel-fuelled types.

464 Ensure that:

- they are of a suitable rated power output for the intended use;
- they have been tested and certified by a competent person;
- they are sited in a well-ventilated place away from LPG cylinders and combustible material;
- they are adequately guarded to avoid accidental contact by people or combustible material;
- cables and sockets are appropriate for their intended use;
- the electrical installation is protected by a residual current device (RCD);
- cables do not create a trip hazard;
- fuelling and re-fuelling are carried out in a safe manner;
- fuel is stored in a safe manner in suitable containers.

Fire-fighting equipment

465 Suitable fire-fighting equipment should be provided at the catering operation dependent on the activity type. The equipment must conform to the relevant British Standard (see chapter on *Fire safety*). No combustible materials should be allowed to accumulate next to any catering outlets.

466 Suitable equipment levels are:

- Non cooking: One x 2 kg dry-powder extinguisher;
- Cooking: One x 2 kg dry-powder extinguisher and a 1 m^2 light-duty fire blanket (BS 6575: 1985) or if deep-fat frying, 9 L foam-type fire extinguisher and a 1 m^2 light-duty fire blanket.

Alcohol and bar areas

467 Alcohol comes under the definition of food and should meet the requirements of the relevant food safety legislation, associated industry guides and codes of practice. Ensure that:

- the structure used for the sale of alcohol, usually marquees or tents, complies with the structural requirements (see chapter on *Structures*);

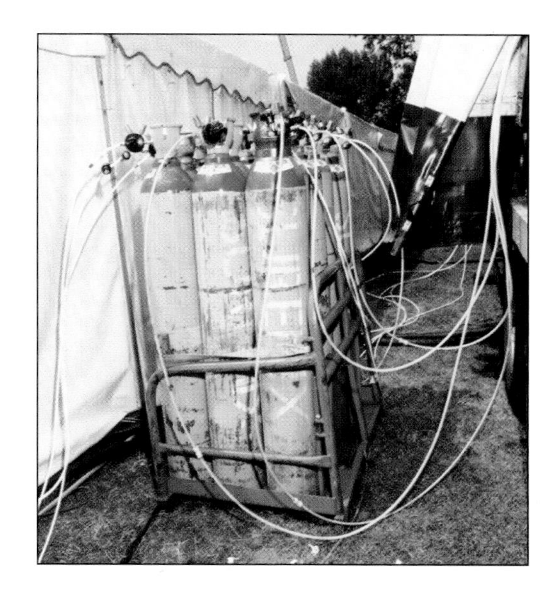

- the operation is designed to allow the free flow of people to and from the bar servery areas to prevent congestion and crushing hazards (this may involve the use of suitable barriers, providing consideration has been given to the barriers becoming a hazard in themselves);
- the electrical installation complies with the requirements detailed in the chapter on *Electrical installations and lighting;*
- suitable and sufficient lighting is provided;
- alcohol storage tanks are positioned on stable, even ground allowing suitable access for delivery vehicles, particularly in bad weather;
- risk assessments for both food and health and safety, have been carried out;
- carbon dioxide cylinders are suitably secured;
- chemicals to clean pipelines are properly handled and stored;
- the type of containers that drinks are served in conform to any site/event specifications, eg no glass policy;
- there is a suitable means of disposal for glass bottles, used to decant drinks before serving;
- bar areas are kept free of litter and the floors are cleared of spillages;
- if a 'token system' is used instead of cash, the 'change areas' need to be separate from the bar service area.

Drinking water

468 The provision of free drinking water is important at all events, especially open-air concerts and dance events, due to the volume of people, confined conditions and the weather.

469 Generally all water should be provided from a mains supply, but if this is not possible then bowsers are permissible provided they are suitable for the purpose. All water dispensing equipment should be clean, well maintained and suitable. It is considered good practice to sample and test temporary water supplies for bacteriological safety, especially those provided at outdoor events.

Pit area

470 There should be an adequate supply of drinking water points in the pit area, together with an adequate supply of paper or plastic cups. The number of drinking water points will be determined by the risk assessment.

471 If storage containers are used to supply the water, they should be of sufficient capacity and number for the anticipated needs of the people within the first 5 m of the pit barrier. Pit area water points should not be within the reach of the audience.

General area

472 There must be a supply of drinking water within easy reach of the audience and all catering operations. At outdoor sites (one-day events) a general guideline is one water outlet per 3000 people and one outlet per ten caterers provided they are in the same area.

473 All water points should:

- have unobstructed access;
- be clearly marked;
- be clearly lit at night if the event continues after dark;
- have self-closing taps.

474 The ground surrounding all water points should be well drained or provision made to 'bridge' any flooded areas.

Chapter 12

Merchandising and special licensing

475 There are four aspects to merchandising that need to be planned and managed:

- the merchandising facilities which include the structure of the stalls or stands;
- the space requirements;
- the setting up, dismantling and operation of the stall or stand;
- the items for sale as merchandising.

Facilities

476 It is essential that merchandising stalls and stands are considered in all aspects of the planning and management of the music event.

477 Consider the following matters when planning the venue or site design:

- the position, size and space requirements of the merchandising stalls or stands within the arena or venue to ensure that entrance and exit audience flows are not obstructed, or cause an audience build-up at any strategic points;

- whether stands and stalls are of a fixed or temporary nature?
- check that any structures will be erected properly and will satisfy any structural integrity requirements (see chapter on *Structures*), as well as requirements in respect of fire safety (see chapter on *Fire safety*);
- power supplies, if required, need to be considered as part of the overall electrical supplies to the event (see chapter on *Electrical installations and lighting*);

- any vehicle or vehicle movements associated with the stands or stalls;
- allocation of parking spaces and camping accommodation for people working at the stalls or stands;
- waste accumulation and collection;
- security arrangements.

478 Ensure that people working on merchandising stalls and stands are informed of the site safety rules particularly in relation to the equipment that can or cannot be brought onto site or within the arena or venue. Also make them aware of the space allocated to them on site and that they must adhere to allotted space.

Setting up, operation and dismantling

479 Most workers employed by people running merchandising stalls and stands are temporary. To ensure that all relevant safety information is passed to all workers, brief the people running the stalls and stands about safety matters on site. Define the responsibilities for health and safety and agree methods of communication with the merchandisers. Give a copy of the site safety rules to the merchandisers when they arrive on site and ensure that they and any subcontractors are informed of the site safety rules.

480 Checks should be made on any public and products liability insurance certificates. Agree the operation time of the merchandising stands with the operator and explain procedures to be taken in the event of a major incident or contingency. Any gas or electrical equipment brought onto site by merchandisers should be accompanied by relevant inspection reports and have undergone the recommended testing. Other equipment should be examined to ensure that the relevant fire-fighting equipment is available in case of fire.

481 In the case of permanent sites, information on health and safety policies within the premises will already be in place. Therefore, the procedures should be followed at all times by all concerned.

482 Stewards working on behalf of the merchandisers, who do not form part of the event stewarding teams, should be approved and involved in the event briefing and lines of communication and co-ordinating activity agreed. Discuss the use of radios for communication to avoid conflicting frequencies.

483 The storage of merchandising stock, particularly if flammable goods are for sale should be discussed with the fire authority and local authority to ensure that the appropriate fire extinguishers are on hand on the stands or stalls. The control and movements of stock around the site should follow agreed procedures.

Items of merchandising

484 The items for sale as merchandising should not breach any licence requirements, trading standards, copyright, or trademark regulations.

485 Ensure that information concerning items that could cause injury or discomfort (eg glo-sticks) is given to the purchaser at the point of sale and the procedures for their correct use are prominently displayed.

486 The practice of tattooing, body piercing and massage will require a special treatment licence from the local authority in certain parts of the UK. Check that the necessary licences have or will be issued by the local authority before allowing tattooing or body piercing to take place.

487 Offensive material should be viewed in relation to the audience profile and perhaps not actively displayed.

488 In the case of ticket touts and unwanted street traders, liaise with the local authority to determine the methods that can be used to deter such practices.

Chapter 13

Amusements, attractions and promotional displays

489 Guidance is already available in relation to attractions, rides, amusement devices and stalls found at fairgrounds and amusement parks (see HSE's publication HSG175 *Fairgrounds and amusement parks: Guidance on safe practice*). This chapter does not replace the need for an event organiser to obtain this guidance. The intention of this chapter is to highlight some areas for consideration when amusements, attractions and displays are incorporated in a music event rather than at a fairground or amusement park.

Amusements and attractions

490 If you wish to include amusement activities at your event, it is important to obtain the required safety information about the activity from the operator. This is to ensure that the siting and operation of the amusement does not:

- compromise safety in relation to the overall risk assessment for the event;
- block the emergency access routes; or
- cause audience congestion problems.

491 Points to consider when incorporating any amusement as part of the overall entertainment include the following.

- Obtain advice about the particular hazards associated with the amusement or attraction from the operator and ask them for copies of their own risk assessment and safety information. Incorporate the information into your overall risk assessment for the event.

- Obtain advice from the relevant enforcement authority (local authority/HSE) about the particular amusement. Local authority officers and HSE inspectors should have up-to-date information concerning hazards that have been reported about a particular amusement activity.
- Check the competence of the operator. It should be relatively straightforward to check the competence of the operator against information already acquired. Is the operator able to demonstrate compliance with legislation or codes of practice? Are they a member of an association? Do they have current insurance? Does each amusement have a current certificate of thorough examination from an inspection body? What experience have they had in operating the amusement? What safety information can they supply in relation to the amusement?
- Information concerned with the safe operation of the amusement should also be given to other contractors working at the event who may be affected.
- Determine appropriate setting-up times, operating times and dismantling times. Amusements should be set up before the audience enters or approaches the event. Make sure that the amusement is not dismantled until all members of the audience have left or are at a safe distance. Vehicle movements are often prohibited during events and amusement operators need to be informed about this policy.
- Ensure that suitable space has been allocated for the amusement. Space is one of the most important considerations for any amusement. This does not just include space on the ground but often space above. Obstacles such as large trees, overhead-cables and power-lines can cause major hazards to the safe operation. The sides and rear of the amusement may need barriers to prevent members of the audience being exposed to hazardous parts of the equipment. The space allocation must therefore be considered in your venue and site design. Minimum space requirements can be found in *Fairgrounds and amusement parks: Guidance on safe practice.*
- When planning the positioning of the amusement, consider emergency access routes as well as space for members of the audience who may be queuing to 'have a go' on the amusement. Space may be needed for family, friends and others to comfortably watch the amusement.
- Ensure that the operation is co-ordinated with the music event. Crowd management problems can arise if operators are still offering rides or 'go's' on the amusement after the music event has ended and if members of the audience try to have one last go before leaving. On the other hand, it may be appropriate to continue the operation of the amusement to stagger people leaving the event. Whatever the decision, careful co-ordination of the activities must be planned and communicated to the operator and stewards.
- The availability of natural light may also be an important safety factor in the operation of some amusements, particularly where colour-dependent safety features are used.

Bungee jumping

492 The British Elastic Rope Sports Association (BERSA) Code of Safe Practice was formulated in consultation with HSE, the Safety and Reliability Directorate of the Atomic Energy Authority and independent operators. The Health and Safety at Work etc Act 1974 (HSW Act), the Management of Health and Safety at Work Regulations 1992 (Management Regulations) and the Provision and Use of Work Equipment Regulations 1998 (PUWER) also apply to bungee jumping. Ensure that bungee jumping operators belong to a reputable association, which can meet the requirements of the BERSA Code of Safe Practice.

493 Pre-booking is recommended for those wishing to take part in bungee jumping and this should be discussed with the bungee jumping operator. In particular, advertising in advance of the event may need to be considered.

Inflatable bouncing devices

494 HSE's *Safe operation of passenger carrying amusement devices: Inflatable bouncing devices* PM76 describes various factors that can contribute to accidents involving inflatable bouncers and the precautions that should be taken to avoid them.

495 There are many types of inflatables that can be used for bouncing upon, including:

- open-sided or flat beds;
- open-fronted, eg castles;
- totally enclosed.

496 Hazards include being blown over or away by the wind, splitting of the fabric, accidental spilling of users, injury to the users by themselves or other users, overcrowding, air loss due to blower disconnection, power supply failure and inadequate means of escape in the event of fire. Each inflatable should be thoroughly examined annually for any deterioration by a competent person or company. Height and age restrictions are often necessary for the safe operation of these bouncing devices and such information should be made visible to the audience wishing to take part.

Flight simulators and computer games

497 Mobile flight simulators and associated rides are often very heavy pieces of equipment that need specific ground conditions to operate safely, access requirements and ample space around the device. They are also defined as fairground equipment and subject to the requirements of *Fairgrounds and amusement parks: Guidance on safe practice.*

Fairgrounds and fairground rides

498 Fairgrounds or individual fairground rides may be incorporated at outdoor events and some larger indoor events. *Fairgrounds and amusement parks: Guidance on safe practice*, provides guidance and should be consulted. The enforcement responsibility of fairgrounds rests with HSE.

499 There is also a series of plant and machinery guidance notes that detail the safety requirements for individual passenger-carrying devices. Current titles available include: *Safe operation of passenger carrying amusement devices; The waltzer; The octopus; The cyclone twist; The big wheel; The paratrooper; The chair-o-plane; The roller coaster; The ark/speedways; The water chute* and *The trabant.* Liaise fully with the operators of these devices to ensure that the requirements laid out in the guidance notes can be fully met.

Circuses

500 It is relatively unusual for a complete circus to be part of a music event. It would be more common to find circus performers demonstrating some of their talents in and around the venue or site itself. Performances and acts such as fire eating, stilt walking and juggling are examples. Brief performers about safety matters in relation to audience safety and advise them where and when to start their act. Emergency exits should always be kept clear. The enforcement responsibility for circuses is

allocated to local authorities, with the exception of a small number of circuses operating within fairgrounds, which remain the responsibility of HSE.

Promotional displays

501 Companies sponsoring events may wish to advertise their product by way of a promotional display. These can range from advertising balloons and inflatables, purpose-made structures, video and virtuality electronic games through to smaller merchandising stands.

502 It is easy to overlook the effect that some of these displays might have on the safety of the event. Obtain information as to the type of equipment that will be brought on site, its method of erection and particular hazards that the equipment may pose. Drawings of any special structures should also be obtained along with the methods of erection and dismantling.

503 Consider the placing of promotional displays at the venue design stage to ensure that they do not obstruct emergency exit routes or hamper audience movement around the site. Inflatable balloons and displays must have appropriate space allocated to them as well as being suitably anchored.

504 Those bringing the equipment onto site must also be instructed on the site safety rules. Any advertising stands should be treated in the same way as merchandising stands. Electrical equipment must come equipped with the relevant electrical test certificates and be installed by a competent electrician (see chapter on *Electrical installations and lighting*).

Chapter 14

Sanitary facilities

505 Ensure that adequate sanitary provision is made for the number of people expected to attend the event, and that consideration is given to location, access, construction, type of temporary facilities, lighting and signage.

506 Construct and locate toilets so that people are protected from bad weather and trip hazards. The floors, ramps and steps of the units should be stable and of a non-slip surface construction. Protect connecting pipe work to avoid damage.

507 Toilets should be readily visible, lit, and clearly signed from all parts of the venue. The areas and, where appropriate, the individual units, should be adequately lit at night and during the day, if required. The Chartered Institute of Building Services recommends a minimum lighting level of 100 lux for general toilet areas (200 lux for wheelchair-accessible toilets).

Maintenance

508 Regularly maintain, repair and service toilets using suitably experienced competent workers throughout the event to ensure that they are kept safe, clean and hygienic. Toilets need to be supplied with toilet paper, in a holder or dispenser at all times. Arrangements should be made for the rapid clearance of any blockages.

Location

509 Where possible, locate toilets at different points around the venue rather than concentrating in one

small area, to minimise crowding and queuing problems. Consider placing toilets outside the perimeter fenced venue area (eg car parks, box office queuing areas, event campsites, etc). Attention should be given to access requirements for servicing and emptying. This may include the need for temporary roadways and dedicated access routes, subject to the layout of the site.

Type

510 Where temporary toilets are required, an assessment should be made of the suitability of each of the available types of temporary unit, for the nature and duration of the event being organised. Consider the perceived peak usage of any toilet units and the time taken for cisterns to fill. Rapid and constant use of any toilet can cause the bowls to become unsanitary and prone to blockages.

511 Temporary mains units can be used if a sewer, drain, septic tank, or cesspool is available, provided an adequate water supply and adequate water pressure are available. Recirculating self-contained units are not reliant on the availability of drains or water services. Provision must be made for servicing vehicles and safe access.

512 Single self-contained units are versatile and easily relocatable during events but are limited to a maximum number of uses before requiring servicing/emptying. Trenches and latrines may be appropriate for some events though advice should be sought from the Environment Agency or Scottish Environmental Protection Agency and local authority regarding their suitability for each event and any local guidelines for ensuring safe and hygienic use.

513 Wherever non-mains units are used, provision for safe and hygienic waste removal must be arranged with holding tank facilities if required. Advice should be sought from the Environment Agency or Scottish Environmental Protection Agency.

Numbers

514 Recommendations as to the minimum scale of toilet provision for buildings of public entertainment are laid out in BS 6465: Part 1 1994. For events licensed for public entertainment, the numbers and location of toilets should be agreed with the local authority.

515 In all circumstances, the sanitary accommodation will depend on the nature of the event, the audience profile, and the type of venue. To calculate sanitary provision requires knowing the audience size and then estimating the anticipated male to female ratio. When there is insufficient information to assess this ratio, a split of male to female 50:50 should be assumed.

516 Consider the following when determining the minimum provision for sanitary conveniences:

- the duration of the event;
- perceived audience food and fluid consumption;
- adequate provision during intervals and breaks in performance;
- requirements for event-related temporary campsites;
- provision of suitable facilities for children, elderly or infirm people attending who may take longer to use a facility;
- facilities inside a fenced venue at a 'no-readmission' event;
- weather conditions and temperature.

517 The experience of a competent consultant or responsible contractor could prove invaluable when determining numbers of sanitary conveniences.

518 The table below shows a general guideline for a music event, though these figures may be too high for short duration/'non peak' period events such as country fairs and garden parties, or too low for events with high levels of fluid consumption or where camping will occur.

For events with a gate opening time of 6 hours or more		**For events with a gate opening time of less than 6-hours duration**	
Female	*Male*	*Female*	*Male*
1 toilet per 100 females	1 toilet per 500 males, plus 1 urinal per 150 males	1 toilet per 120 females	1 toilet per 600 males, plus 1 urinal per 175 males

Washing facilities

519 Where possible, provide hand-washing facilities in the ratio of one per five toilets with no less than one hand-washing facility per ten toilets provided. Provide suitable hand-drying facilities. If paper towels are supplied, arrange for regular disposal and restocking.

520 Where warm water hand-washing facilities are available, provide adequate supplies of suitable soap. Antiseptic hand wipes or bactericidal soap should be provided where warm water is not available.

521 On sites where hand-washing facilities are supplied in the open air, consider the management of the facility to ensure that the surrounding ground does not become waterlogged leading to localised flooding.

522 Control of Substances Hazardous to Health Regulations 1999 (COSHH) assessments should be available to cover all cleaning and deodorising products used. Where products are known to present a risk to users with pre-existing skin conditions, suitable warning notices should be prominently displayed.

Long duration events

523 Hand-washing facilities alone may not provide adequate provision for events longer than one day, or when overnight camping is available. In these instances, consider whether it may be appropriate to supply shower facilities on site, subject to the availability of adequate water supply and water pressure.

Sanitary provision for people with special needs

524 Provide appropriate sanitary accommodation for wheelchair users and other people with special needs attending the event. The Disability Discrimination Act 1995 will apply with regard to sanitary accommodation for people with special needs.

525 Also consider access to toilets for people with special needs. Supply fixed and stable ramps where appropriate. Position facilities close to any area set aside for people with special needs such as viewing platforms, and ensure they are designed to comply with the provisions of BS 5810: 1979.

526 The provision of facilities should relate to the expected numbers of people with special needs attending the event. It is suggested that one toilet with hand-washing facilities should be provided per 75 people with special needs.

Disposal of sanitary towels and nappies

527 If there is any possibility that sanitary towels or nappies may block sanitary conveniences, supply suitable and clearly identified designated containers with suitable arrangements for regular emptying of the receptacles.

528 If infants are expected at an event, provide appropriate baby-changing facilities including receptacles for the hygienic disposal of nappies. Provide prominent signs within the baby-changing cubicle to ensure that the receptacles are used.

Sewage disposal

529 Different water authorities have different policies regarding waste disposal, and many disposal sites are closed at night-times and at weekends. If effluent needs to be stored on site until off-site disposal facilities are open, it is essential that adequate holding tanks are provided on site in a safe and secure location. Seek advice on safe effluent disposal from the appropriate water authority and local authority and ensure that a licensed contractor is employed to remove and dispose of effluent. Arrangements should be documented and agreed with the contractor before the beginning of the event.

Facilities for employees and event workers

530 The Workplace (Health, Safety, and Welfare) Regulations 1992 require that suitable and sufficient toilets and washing facilities must be provided at workplaces. Guidance on the facilities that should be provided is given in the accompanying code of practice to the regulations L24 *Workplace health, safety and welfare*.

531 Sanitary accommodation for use by event workers, should be located near to the work areas and, in particular, behind the stage, near the mixer tower, next to the catering areas and car parks, the first-aid areas, welfare and children's areas. Specific toilets with hot and cold hand-washing facilities should be provided for food handlers.

Contractors providing or servicing the sanitary facilities

532 Discuss requirements for the type, numbers, positioning, servicing and maintenance of sanitary facilities with the contractor before the event. It is advisable to provide contractors with a plan of the site, showing the proposed location of the facilities along with a copy of the site safety rules and information concerning any significant risks highlighted in the overall event risk assessment. Separate waste transfer notes will be required for containers holding waste from sanitary towels and nappies.

533 Examine contractors' safety policies and risk assessments. Contractors should ensure that their workers are provided with and wearing the correct personal protective equipment. Protective overalls, boots or shoes, gloves and eye protection are needed to ensure that workers are protected from accidental splashes of the disinfecting and odorising chemicals as well as accidental contamination by sewage.

534 An assessment is required under the COSHH Regulations by contractors providing, servicing, or emptying the sanitary facilities. The COSHH assessment needs to consider exposure to the chemicals used in the recirculating self-contained units as well as accidental exposure to sewage.

Chapter 15

Waste management

535 Large quantities of waste materials will be generated by the concessionaires and the audience at most music events. Waste needs to be managed carefully to minimise the risks associated with its accumulation, collection and final disposal.

Types of waste

536 Types of waste generated include the following:

- paper and cardboard packaging;
- food and drink containers;
- left-over food debris;
- waste food from food concessions;
- glass;
- plastics;
- metal cans;
- other metal waste, eg construction materials;
- clothing;
- human waste products (vomit, urine and faeces, sanitary towels and tampons often placed in miscellaneous containers);
- medical waste such as needles and bandages;
- remains of camp fires;
- fireworks and pyrotechnics;
- waste water from toilets; showers and hand-washing basins;

- waste water from food concessions;
- needles used by IV drug users.

Hazards posed by waste

537 Hazards posed by waste include the following:

- injury to workers during collection and removal of waste from the site. Examples include cuts and grazes, needle stick injuries; back strains due to manual handling difficulties and possible infection;
- accumulations of waste, blocking emergency access routes or hampering movement around the site as well as presenting tripping hazards to the audience;
- fire hazards when waste is accidentally or purposely ignited;
- the misuse of waste by the audience, eg throwing bottles, cans, etc;
- vehicle movements associated with the collection of waste materials;
- waste attracting insects and vermin.

Areas where waste is generated and the types of waste

538 Waste and the type of waste products will not be generated evenly across the venue or site. The build-up of waste will vary in different areas over time. A competent waste contractor will therefore need to manage their workers and equipment to ensure that there are suitable and adequate resources directed to the most appropriate areas at the most appropriate times. Each area of the venue or site may need to be managed differently.

539 Pay special attention to the following areas:

- approaches to music event, ie surrounding streets or land;
- entrances and exits;
- arenas and stages;
- sanitary areas;
- first-aid areas and health-care waste;
- catering areas;
- camping area.

Information to be exchanged with waste contractor

540 Ensure that details are given to the waste contractor concerning audience size, arena size, site boundaries, numbers of campers, food concessions, and other relevant factors. The waste contractor cannot accurately plan working methods or employ the correct number of workers without this information. Insufficient information could have serious consequences for the audience and employee health, safety and welfare.

Methods of collection

541 The collection of waste from the site or arena usually involves a combination of the following:

- contractors' workers specifically trained to pick the waste up (litter pickers), and/or empty the receptacles placed around the site or venue;
- the use of sweeper vehicles and vacuum suction vehicles;

- vacuum tankers for collection of waste water temporarily held in smaller tanks;
- other vehicles, trailers and towing vehicles.

542 Discuss arrangements with the waste contractor before the event so that any special requirements regarding access or height restrictions, storage space for vehicles or accommodation for the litter pickers can be incorporated into your overall event planning.

Receptacles

543 Waste receptacles can be positioned around the perimeter of the venue or site, and they can also be positioned within the venue or site or other areas as appropriate. Great care must be exercised in choice, size and location of receptacles. Wheeled containers or similar receptacles appear to be the most versatile at present as they can be obtained in a variety of sizes and are equipped with lids. They can be easily positioned and manoeuvred as required. Also consider providing tamper-proof sharps bins.

544 Steel drums are difficult to manoeuvre and empty when full so assess their use. Skips can also be used but again their positioning requires planning to ensure that there is suitable access for delivery and collection especially in wet weather. Generally, position them in areas separated from the audience. Waste receptacles can be set on fire, so require regular monitoring.

545 Large on-site compactors can also be used to reduce the bulk of the refuse. They will need to be plugged into an electrical supply and a trained operative should be available at all times. Front-end-loader containers may also need to be separated from the audience for reasons of safety, access for loading and to prevent the audience placing uncompactable and hazardous or other inappropriate waste into these units.

546 The collection company must be a registered waste carrier or exempt from registration. Vehicles used to help with the collection of waste must be mechanically sound and be accompanied with the relevant test certificates including an MOT if appropriate.

Times of collection

547 Discuss with the waste contractor the strategy for collection of the waste for the whole duration of the music event, including pre- and post-event collections. Different collection methods may need to be planned for each of these phases.

Methods of removal

548 Discuss with the waste contractor the methods of removal of waste from the venue or site. There

may be areas that are subject to a ban on vehicle movements during the event to protect the audience. The sites chosen for the bulk collection must have a suitable access route capable of taking the weight of often very large collection vehicles. These could weigh up to 38 tonnes.

549 An event organiser is under a 'duty of care' discussed in paragraphs 557-559 . The responsibility for the disposal of waste does not end with the waste leaving the site or venue but extends until the waste has reached its final destination.

Health, safety and welfare of employees and event workers

550 Waste contractors have a legal duty to ensure that the health, safety and welfare of their employees is protected on site. The Personal Protective Equipment Regulations 1992 require an employer to ensure that their employees have suitable and sufficient protective clothing and equipment to carry out their tasks.

551 Examples of suitable clothing include:

- protective boots or shoes with metal toe caps;
- trousers and jackets;
- waterproof suits;
- fluorescent waistcoats;
- hard hats;
- goggles;
- different types of gloves for different tasks.

552 Ensure that toilets are available throughout the waste collection process. Those handling waste need access to hot and cold running water, soap and nail brushes to wash their hands and bodies if they become contaminated. Toilets and washing facilities must be available, particularly at the final waste collection process and in some circumstances showers will be necessary.

553 Brief workers before beginning work to explain site hazards and risks, hours of work and meal breaks and estimated completion time.

Recycling

554 The options for recycling on site include designated recycling containers around the site. These can either be small-wheeled containers, which are then collected to allow further sorting of the materials to take place, or they can be larger 'banks' such as those provided by specialist contractors. The effectiveness of the segregation systems for recycling will be dependent upon the good will of the attendees to the event, adequate provision of the recycling containers, suitable clear labelling of the containers and the location of the containers.

Legislation

555 Waste holders (which includes waste producers, waste carriers and waste disposers) have a duty under the Health and Safety at Work etc Act 1974 (HSW Act) to ensure, so far as is reasonably practicable, the health and safety of their employees and other people who may be affected by their actions in connection with the use, handling, storage or transport of waste. A waste producer must also ensure that a safe means of access to and exit from the waste storage area is provided for the contractor. Event organisers will be defined as waste producers.

556 Legislation concerning the management of the waste itself includes the following:

- the Environmental Protection Act 1990;
- the Environment Act 1995;

- the Controlled Waste Regulations 1992;
- Waste Management Licensing Regulations 1994;
- Special Waste Regulations 1996.

The duty of care

557 Section 34 of the Environmental Protection Act 1990 introduced a duty of care for waste management which applies to anyone who produces, carries, treats or disposes of controlled waste. Controlled waste is defined as any commercial, industrial or household waste.

558 A waste producer is under a legal obligation to ensure that the waste is only collected by a registered waste carrier or someone registered as exempt such as a local authority. It should only be transferred to a site with a suitable waste management licence. The waste must be accompanied by an accurate description of the waste and transfer notices must be completed and signed. The duty of care also imposes a duty for people to contain the waste on site.

559 It is an offence to deposit controlled waste on land not in accordance with a waste management licence and also an offence to keep, treat or dispose of controlled waste in a manner which is likely to lead to pollution of the environment or harm to human health. It therefore follows that the waste producer, in this case the event organiser, must ensure the selection of competent and responsible transport and disposal contractors. The waste producer must ensure that whoever has possession or control of the waste will handle and eventually dispose of it legally. Obtain the Environmental Protection Act 1990, section 34: *Duty of care: A Code of Practice* for further information.

Chapter 16

Sound: noise and vibration

560 High sound levels present a risk to hearing, both for those working at an event and for the audience. High levels of vibration can have serious consequences for the integrity of temporary and permanent structures. Both sound and vibration can lead to noise nuisance outside the venue. Therefore, proper control and management of sound and vibration levels is needed both in rehearsal and during the event.

561 If sufficiently loud, any sound, including music, can damage hearing if people are exposed to it long enough. The risk to hearing from loud sounds is directly related to the dose of sound energy a person is exposed to. The risk of damage to hearing increases the louder the sound and the longer a person is exposed to it. At high sound levels the risk of damage to hearing occurs at much shorter exposure times than at a lower levels; at extreme high or impulsive levels the risk of injury to the ear is almost immediate.

562 Most members of the audience will not attend events regularly enough to suffer serious hearing damage solely as a result of going to music events. However, the louder events can contribute significantly to the overall sound exposure that members of the audience receive throughout their life, including noise from other leisure activities, at work and at home, therefore increasing the risk of damage to their hearing.

563 The Health and Safety at Work etc Act 1974 (HSW Act) and the Noise at Work Regulations 1989 require you to protect workers and the audience from noise. The Management of Health and Safety at Work Regulations 1992 (Management Regulations) also apply to cover noise and vibration considerations.

564 For the community impact of noise from events, many local authorities already have environmental music noise control protocols which they apply to venues in their district. The Noise Council has produced a *Code of practice on environmental noise control at concerts* and recommends noise control procedures for minimising noise in surrounding areas. Refer to this source of guidance for the control of environmental music noise and its impact on communities neighbouring outdoor music events.

565 In terms of vibration impact, the effects off site will generally be much less significant than on site, with the nuisance aspect of vibration being most significant. For the potential nuisance aspects of vibration, guidance is available in BS 6472: 1992.

Workers

566 The Noise at Work Regulations 1989 gives the legal duties (on an employer) to prevent damage to the hearing of workers from excessive noise at work. They set out actions which must be taken when stated levels of noise exposure are reached.

567 If noise exposure is likely to reach the first action level of Lep,d of 85 dB (A) (Lep,d = daily personal noise exposure level) or peak action levels, employers must:

- ensure that a noise assessment is made by a competent person;
- provide workers with information and training;
- provide ear protection for all workers who request it.

568 If noise exposure is likely to reach the second action level of Lep,d of 90 dB (A) or the peak action level of 200 pascals (140 dB), employers must:

- ensure that a noise assessment is made by a competent person;
- provide workers with information and training;
- reduce exposure as far as is reasonably practicable by reducing sound levels or the time exposed to the noise or both (without ear protection);
- provide ear protection to all workers and ensure that they are used correctly. The Regulations also require workers to comply with the employer's instructions in respect of noise exposure, including wearing ear protection or taking breaks where necessary;
- mark ear protection zones and make sure that everyone who goes into them uses ear protection. This can include entrances to the stage, monitor mixing area, front barrier area, front-of-house sound mixing and lighting towers, and delay/distribution loudspeaker towers.

569 An essential element of the above is the assessment of noise exposure by a competent person. Guidance on how to choose a competent person and how to carry out a noise assessment can be found in the HSE publication L108 *Reducing noise at work*.

570 Consider commissioning a noise-exposure risk assessment with recommendations for limiting exposure for workers and audience, before organising an event. These noise-exposure risk assessments can then be shown to local authorities before the event takes place and help prevent difficulties occurring on the day of the event. These noise-exposure risk assessments can also help satisfy the requirements of the Management Regulations.

571 The use of regular audiological testing of hearing, for people who are likely to routinely work in high-noise-level environments is recommended by the British Audiological Association and is a requirement of the Management Regulations. Regular audiometry testing allows hearing damage or susceptibility to hearing damage (individual sensitivity to noise-induced hearing loss can vary significantly) to be picked up. See HSE's leaflet *Health surveillance in noisy industries* and guidance note MS26 *A guide to audiometric testing programmes.*

Audience

572 There is no specific legislation setting noise limits for the audience exposure to noise. However, the general requirements of the HSW Act and civil law duties relating to negligence reveal that audiences need to be protected against and informed of the risk of damage to their hearing.

573 The event equivalent continuous sound level (Event Leq) in any part of the audience area should not exceed 107 dB (A), and the peak sound pressure level should not exceed 140 dB

574 The above sound-level exposure values are for the whole of the audience area. For practical purposes, it is usual for audience sound-level exposure to be monitored close to the front-of-house sound mixing position. For the largest outdoor and indoor venues, this can be up to 75 m from the front-of-stage barrier position where the audience sound-level exposure can be significantly higher than at the front-of-house sound mixing position.

575 Ensure that during the sound check the difference in sound level between the front-of-house sound mixing position and the front-of-stage barrier, and where delay/distribution stacks are in use, at the barrier for each delay/distribution stack, is established. This will then allow a guideline sound pressure level for the front-of-house sound mixing position to be determined which will restrict the whole of the audience sound-level exposure to below an Event Leq of 107 dB (A), and peak sound pressure levels to below 140 dB.

576 Where practicable, the audience should not be allowed within 3 m of any loudspeaker. This can be achieved by the use of approved safety barriers and dedicated stewards, wearing appropriate ear protection. Where this is not practical, the overall music sound levels will have to be modified so that people closer than 3 m to the loudspeakers are not exposed to an Event Leq of more than 107 dB (A) or peak sound pressure levels of more than 140 dB. Under no circumstances should the audience and loudspeaker separation distance be less than 1 m.

577 Where the Event Leq is likely to exceed 96 dB (A) advise the audience of the risk to their hearing in advance, eg either on tickets, advertising or notices at entry points.

578 Sources of noise other than music also need to be properly controlled. In particular, the noise from pyrotechnics should be restricted so that at head height in the audience area, noise from pyrotechnics does not exceed peak sound pressure level 140 dB or 200 Pa. Discuss this requirement with the specialist pyrotechnic technicians before the event, as charge density and altitude of deployment may need adjusting to meet this requirement.

579 Noise sources such as music associated with funfairs and sound systems brought onto site by merchandising concessions can also add to the overall noise levels produced by the event as a whole. Also consider an assessment and control of these sources.

Noise assessments

580 To enable effective management of sound and vibration levels, both in terms of ear protection and external nuisance to the nearby community, a pre-event assessment of likely sound levels, coupled with monitoring and control of sound levels during the event will be necessary. Assessment of

audience noise exposure and off-site noise nuisance levels is best combined with the noise assessment for workers required by the Noise at Work Regulations 1989.

581 This assessment should include the following:

- the sound levels likely in the audience area;
- the steps necessary to ensure that the sound levels likely in the audience area are within the values mentioned in paragraph 573;
- if the sound levels likely in the audience area are expected to exceed the value in paragraph 577, then advance warnings for the audience will be necessary;
- if any worker's noise exposure reaches or exceeds any of the action levels set out in the Noise at Work Regulations 1989;
- if the marking of ear protection zones is necessary, identify the location of such areas;
- the arrangements for monitoring and control of sound levels during the event;
- the sound levels likely outside the venue;
- the positioning, array, type and specification of the loudspeakers making up the sound system. The design, composition, array and positioning of the sound systems used can have very significant benefits in aiding the control and management of noise levels. This is both in terms of the music noise and vibration levels at the venue and outside in the nearby community;
- the possibility of sound and vibration energy being transmitted through the staging, ground and structures, particularly bass and sub-bass sound and vibration energy.

582 Sound and vibration energy at the low bass and sub-bass frequencies has the potential to compromise the integrity of structures, in particular relatively light-weight temporary structures such as staging, lighting rigs, scaffolding platforms, giant video screens and temporary audience stands. The assessment should include an analysis of the risks associated with low bass and sub-bass sound and vibration including reference to BS 7385: Part 1 1990 and Part 2 1993 regarding the measurement of vibration and evaluation of effects in buildings, including ground-borne vibration.

583 Where the assessment identifies potentially hazardous levels of low bass and sub-bass sound and vibration, it should include recommendations for reducing these to acceptable levels and the monitoring of low bass and sub-bass sound and vibration during rehearsal, sound check and performance phases of the event.

Controlling sound and vibration levels

584 Measures to control worker exposure include:

- restricting the length of time spent in noisy conditions;
- restricting music noise levels during rehearsals and sound checks;
- shielding of work areas where workers do not need to fully hear the music to function properly, ie backstage, under the stage, stage wings, in artist and guest hospitality areas, first-aid posts, and in areas used for organising the event, safety, control and administration, etc;
- using ear protection in areas which are likely to exceed the first action level of Lep,d of 85 dB (A) and where there is no other practicable way of reducing the noise level below this level.

585 Ear protection should ensure that when worn properly the sound exposure of the person is below the second action level or peak action level of the Regulations. The Regulations require that where ear protection is provided, workers must wear it in the appropriate ear protection zones.

586 Ear protection should be provided by the employer. It should be allocated to appropriate workers at risk as identified in the event risk assessment. Additionally, ear protection should also be provided as back-up near to the ear protection zones identified in the event risk assessment. If communal stores or dispensers of ear protection equipment are used, the equipment should be checked for wear and tear and restocked before the start of each event and during the course of each event as necessary.

587 If self-employed people are on site and are likely to be exposed to noise levels which exceed the first action level of the Noise at Work Regulations 1989 (Lep, d 85 dBA), notify them in writing to comply with the noise exposure control regime for the event, stating that they must provide their own ear protection.

588 For certain workers such as sound engineers and the performers, it is often argued that they need to be able to hear the music at least at the same level as the audience. This frequently results in these people being routinely and repeatedly exposed to noise levels which exceed the first action level of the Noise at Work Regulations 1989 (Lep, d 85 dBA), with accompanying serious risk of hearing damage.

589 Ear protection, both passive and active, is now available which equally reduces sound across the frequency range. It has been successfully used by sound engineers and musicians, etc, to restrict their exposure to high music noise levels, without unacceptably compromising the quality of the music sound they hear. Encourage relevant workers to use such specialist ear protection.

590 Sound levels often rise during an event to maintain impact or to emphasise leading performers who appear towards the end of a show. Where this is likely to occur, the sound level for earlier performers should be set lower to allow for the likely increase, so that the overall sound-level limits are not exceeded.

591 Where control of the sound system is to be transferred to another engineer during the course of an event, all the engineers involved need to be informed about the sound-level monitoring and control system.

Monitoring sound and vibration levels

592 Monitoring of sound and vibration levels during rehearsal, sound check and performance elements of an event is necessary so that there is adequate control of sound and vibration levels.

593 Monitoring can be continuous or for a succession of short periods, eg up to 15 minutes, to enable the overall noise levels for the event to be established. Noise levels for the audience should be checked at head height in the loudest part of the arena, usually at the front-of-stage barrier. As stated previously if the sound level is measured elsewhere, eg at the front-of-house mixing desk position, a correction needs to be estimated during the initial assessment and applied to allow for the difference between that measurement position and the loudest area.

594 During the event, those involved in monitoring and controlling sound and vibration levels need to be able to maintain a dialogue. If monitoring indicates that the sound and vibration levels are liable to exceed the relevant limits, the sound engineer needs to be advised to adjust the system immediately. All sound engineers need to be instructed to act on the advice of the nominated person responsible for overall control of sound and vibration levels.

Chapter 17

Special effects, fireworks and pyrotechnics

Fog and vapour effects

595 In the context of special effects, the term 'smoke' is often used colloquially to mean fog. Strictly, smoke consists of solid particles suspended in air produced by combustion, whereas fog is composed of liquid droplets suspended in air. Fog can be produced by a variety of processes, but not by burning. Only fog-making processes are considered in this chapter. The three main fog-making technologies are heated fogs, cryogenic fogs and mechanical fogs.

Heated fogs

596 When using heated fog machines, always consult the manufacturer's instructions to ensure that the correct fog fluid is used in the equipment. The relationship between fog fluid composition, temperature control and the design of the fog machine is critical. Use of an incorrect fog fluid may result in either under-heating, which can leave residues, or overheating, which may cause the fog fluid to decompose and produce harmful by-products. Never use mineral oil in a pump-propelled glycol fog machine, as this will create a fire hazard.

597 Carefully follow the manufacturer's instructions for maintenance and cleaning of heated fog machines. Do not modify or bypass the fog machine thermostat and do not use contaminated fog fluid. The composition of the fog fluid should not be altered by the addition of pigments or dyes to change the colour of the fog. When heated, the added substances may make the fog unsafe to breathe, or clog or damage the machine. The use of coloured filters on the lights illuminating the fog will produce coloured fog. Perfumes and scents should not be added to the fog fluid without the manufacturer's approval.

Cryogenic fogs

598 Handle all cryogens with care and follow the manufacturer's safety recommendations as they are extremely cold. Dry ice (solid carbon dioxide) should only be handled when wearing impervious well-insulated gloves of an approved design, as even momentary skin contact causes serious frostbite and blisters. Medium-term storage of dry ice is possible in containers with good insulation, but these should be vented and positioned in well-ventilated areas. As small pieces of dry ice vaporise rapidly, it is advisable not to break up blocks of dry ice until immediately before use.

599 Take special care when handling liquid nitrogen as it is extremely cold and causes severe frostbite on contact with the skin. When handling liquid nitrogen wear long insulated gloves and face visors of approved design. This ensures that no skin is left uncovered to avoid hazards from splashing. Store liquid nitrogen in the container in which it is supplied. Like all cryogens, stored liquid nitrogen should be allowed to vent. Failure to do so will lead to rupture of the container. Consult suppliers for further information and advice on the storage, handling and use of cryogens.

600 Because of their low temperatures, all cryogenically generated fogs tend to stay at low level. In addition, the carbon dioxide in fog produced by dry ice is heavier than air and will quickly move down to the lowest accessible level. However, cryogenic burst techniques can be used to create fog effects high in the air.

601 Carbon dioxide and nitrogen gases cause asphyxia and high concentrations can present dangers to the audience, performers and stage workers. Good ventilation is therefore necessary. Particular care needs to be taken at indoor venues in the orchestra pit, and in any other confined spaces such as under-stage basement workshop and storage areas. The gases may flow into such areas through openings and crevices and put people at risk. In addition, no one should be allowed to lie down in a fog that has been produced cryogenically.

602 Following initial generation of the fog, the vapours become invisible and the concentration of the gas may be difficult to determine. If, during a test before its use at a performance, there is any doubt about the concentration present, you should seek expert advice to monitor the oxygen and carbon dioxide levels before the equipment is used for a performance.

Mechanical fogs

603 Three methods are employed for producing mechanical fogs: pressurised water, crackers and ultrasonic. All of these methods use a mechanical process to manipulate a fluid to produce fog without the need for heating or cooling.

General requirements

604 Guidance on the use of smoke and fog effects in entertainment is given in HSE's *Smoke and vapour effects used in entertainment*. Where adverse health effects are possible, work with smoke and fog effects will be subject to the Control of Substances Hazardous to Health Regulations 1999 (COSHH). A risk assessment should be carried out on the substances used to produce fog and on the composition of the fog itself. Any identified risks must be removed or controlled as far as is reasonably possible. Suppliers and manufacturers should provide the required information to allow the risk assessment to be carried out.

605 Some substances used to create fog effects have occupational exposure limits (OELS) published in HSE's guidance note EH40 *Occupational exposure limits 1999*. These limits should be referred to in the manufacturer's information. Exposure below these levels should cause no ill effects in most people, although small children, the elderly and asthmatics may be at risk.

606 In addition to the COSHH risk assessment covering chemical hazards, a general risk assessment should be carried out in accordance with the Management Regulations. The assessments should consider all people likely to be affected, including fog machine operators, performers, members of the audience and other workers. Particular attention should be given to those who are more likely to be affected such as small children, the elderly, asthmatics and others with respiratory sensitivity. If there is any doubt about the level of exposure that may result from a particular fog effect, on-site monitoring should always be employed.

607 When fog is to be used in a performance, display warning notices or print warnings on the tickets. If necessary, these warnings can be reinforced by verbal warnings before the fog is used.

608 Locate the fog machine in a fixed position and ensure it is adequately protected against unauthorised interference. The fog machine should be operated by a competent operator at all times. The fog machine outlets should always be within the direct view of the operator.

609 Some components in a fog machine can get very hot. Normal fire-safety precautions should be observed. The machine should be adequately ventilated and be readily accessible in an emergency. Where a fog machine is used there should be adequate ventilation of the intended fogged area. The amount of fog in the areas to which the audience is admitted should be limited to the minimum necessary for the desired effect.

610 Where necessary use fans to direct the fog into the desired area to prevent clouding at the point of discharge and possible overspill into other parts of the venue. Do not discharge fog or allow it to drift into exits, exitways, stairways, escape routes, etc, or allow it to obscure exit signs or fire-protection equipment.

611 Before approving the use of a smoke or vapour effect, the local authority and fire authority will consider the presence of any automatic fire detection or fire sensor system installed in the venue, because of the possibility of inadvertent triggering of the system.

612 The use of fog machines can lead to a progressive build-up of slippery residues. This can normally be seen on metal surfaces and sealed concrete floors. However, particular care should also be taken to ensure that the residues do not build up on trusses, catwalks, stairways and similar areas to create a slipping hazard. The residues may also be drawn into the air filters of electronic equipment and clog them.

613 Health and safety inspectors will require the production of documentary evidence of the non-toxicity and non-flammability of the fog, unless the type of equipment is in common use.

Strobe lights

614 Carefully consider the use of strobe lights, as under some conditions they may induce epilepsy in flicker-sensitive individuals. Whenever strobe lights are used, arrange for a prior warning to be given at the entrance to the event or in the programme.

615 If strobe lights are used, keep flicker rates at or below four flashes per second. Below this rate it is estimated that only 5% of the flicker-sensitive population will be at risk of an attack. This flicker rate only applies to the overall output of any group of lights in direct view, but where more than one strobe light is used the flashes should be synchronised.

616 To reduce risk further, mount lights as high above head height as is practicable. Where possible, the lights should be bounced off walls and ceilings or diffused by other means so that glare is reduced. They should not be used in corridors or on stairs. Continuous operation of strobe lighting for long periods should be avoided. Further information can be found in the HSE HELA guidance note *Disco lights and flicker sensitive epilepsy*.

Lasers

617 Laser use in entertainment applications has increased in recent years. They are now commonplace at concerts, night-clubs, exhibitions and at outdoor events. Most of these activities involve the use of laser products that can cause eye injury and sometimes skin burns if used improperly. It is therefore important for manufacturers, suppliers, installers and users to properly assess exposure risks that may arise from the use of these devices.

618 HSE has published comprehensive guidance on the use of lasers in these activities: *Radiation safety of lasers used for display purposes* (HSG95), and explanatory leaflet *Controlling the radiation safety of display laser installations* (INDG224). The guidance describes legal duties and gives practical advice about display laser safety assessment. Companies that manufacture, supply or install display laser installations and those that use them, eg venue operators, should ensure that they are familiar with the content of this document. Ensure that you notify the local authority and health and safety inspectors if you intend to use lasers at your event. A laser safety officer will need to be appointed.

619 Display laser safety assurance is primarily about hazardous beam identification and personal access restriction; *Radiation safety of lasers used for display purposes* gives practical guidance on how this can be achieved. In circumstances where access restriction is not possible, eg during deliberate audience scanning, safety assurance becomes more an issue of controlling beam hazard so that applicable exposure limits are not exceeded. The height of the scanning beams is particularly important, eg above audiences on a dance platform. A minimum height of 3 m above the highest dance platform is recommended.

620 Operators of lasers need to be able to assess display modes, either by measurement or calculation, so that emission levels can be kept below applicable limits. These are published in the BS EN 60825: Part 1 1994 (amd 2 1997) as maximum permissible exposure levels (MPEs); those relevant to display laser safety assessment are included in *Radiation safety of lasers used for display purposes*.

621 MPEs are not statutory limits but they are based upon good international consensus on tissue damage thresholds and are therefore considered by HSE to be authoritative for the purposes of safety assurance and enforcement.

High-power (scenic) projectors

Xenon and HMI lamp systems

622 The operator should be competent in the special handling procedures for xenon and HMI lamps. A xenon lamp burst is possible when cold and therefore gauntlets covering wrist arteries, chest protection and a full face visor covering neck arteries should be worn while handling the lamps. Xenon and HMI lamps reach high pressures when energised, eg 30 bar. Both types of lamp reach high temperatures when energised, eg 95 °C. A lamp burst is possible in either lamp type when energised, with the resultant danger from flying glass, burns and fire. When xenon lamps are being installed, other workers should be warned of the dangers and asked to vacate the vicinity for the few minutes that it takes to install the lamps. The lamp houses of commercial high-power projectors are designed to withstand a lamp burst and to contain the glass within the housing even if the burst happens when energised and they should pose no danger to the audience if installed and operated properly.

623 Xenon lamps and HMI lamps produce significant amounts of ultraviolet (UV) radiation. Commercial high-power projection systems are designed to contain this UV radiation within the projector and so should pose no risk to people if used properly.

624 The arc of xenon and HMI lamps is very bright and housings are designed so that the arc cannot be viewed directly by the operator. Care should be taken to ensure that people are not put at risk by 'blinding' them with the light, especially if they are moving around in otherwise dark environments (eg while entering or leaving a venue).

General requirements

625 All types of high-power projection systems require significant amounts of electrical power. Typically, 32 A three-phase power for xenon lamps and 32-63 A single-phase power for HMI lamps per projector. Electrical systems that are installed for high-power projectors should take this into account and cabling should be rated accordingly (see chapter on *Electrical installations and lighting*). Sufficient dry powder or carbon dioxide fire extinguishers should be provided to ensure coverage of all the areas that house scenic projectors.

626 Projection towers should conform to safe working conditions and careful enquiry should be made into the weight and dimensions of the chosen projection system. Not only to house the projector, but also to allow sufficient working space around the system (see chapter on *Structures*). Projectors being used externally need to be housed in a weather-proof projection structure. Water must not enter the projector while in operation. Projectors need to be protected against unauthorised interference, and staffed or readily accessible by a competent technician at all times when in use.

Ultraviolet light

627 Ensure that lamps are used correctly to restrict exposure to UV radiation and in particular to UVB radiation. To remove UVB radiation, some lamps have a double skin whereas other manufacturers provide lamp housings which have separate filters. Lamps should not be used if the outer skin is broken or if the housing filter is not in place.

628 UV lamp operators need to know the emission characteristics of the lamps so that applicable exposure limits are not exceeded. When replacing lamps or other components which could affect the radiation output, it is important that the manufacturer's advice is followed; it is possible to install incorrect replacement lamps unless the manufacturer's lamp specification is known, eg germicidal units could be used in an entertainment application by mistake.

629 A risk assessment should be carried out for the use of UV radiation at events which should take account of exposure to the audience, performers and workers, particularly in relation to photosensitive reactions.

Fireworks

630 A simple definition of the difference between outdoor firework displays and pyrotechnic stage displays is that firework displays are generally intended as entertainment in their own right; pyrotechnic stage displays are usually used to enhance a particular scene or song, or to draw the audience's attention to or from a part of the stage set.

Regulation and controls

631 The primary control on the supply and acquisition of fireworks are the Fireworks (Safety) Regulations 1997. Under these Regulations only certain defined fireworks within BS categories 1, 2 and 3 may be supplied to the public. Category 4 and larger category 2 and 3 fireworks are prohibited to the public and may only be supplied to specified types of people which include a professional organiser or operator of firework displays. The minimum age for the acquisition of fireworks is 18 years.

Authorisation and product safety

632 Authorisation is required by the Explosives Act 1875 (EA75), as amended by the Placing on the Market and Supervision of Transfers of Explosives Regulations 1993. Order in Council Nos, 6, 6A and 16 prohibit the keeping of unauthorised explosives in licensed stores or registered premises. Authorisation is undertaken by the HSE at the same time as classification (required by the Classification and Labelling of Explosives Regulations 1983) and a competent authority document (CAD) issued. The effect of the classification and authorisation requirements is that only explosives listed on such a document may be manufactured, kept, transported or supplied. HSE has published the *Conditions for the authorisation of explosives in Great Britain* (HSG114).

633 The reference standard for fireworks is BS 7114: Parts 1-3 1988. In the case of category 4 fireworks, which have no specific requirements in the British Standard, the requirements of the conditions for authorisation should be met.

634 Obtain fireworks from suppliers who can demonstrate that their products have been authorised and classified. The supplier should provide detailed instructions and guidance with the fireworks and be able to offer advice and back-up in the event of problems arising.

Keeping of fireworks

635 The EA75, as modified by the Control of Explosives Regulations 1991 requires that explosives are kept in a legal place of keeping. For further information contact the local authority or HSE.

Transport to site

636 The carriage of fireworks and other explosives by road is subject to the Carriage of Explosives by Road Regulations 1996 and the Carriage of Dangerous Goods by Road (Driver Training) Regulations 1996. Both sets of Regulations are amended by The carriage of Dangerous Goods (Amendment) Regulations 1999. These address a wide range of operational issues including selection of vehicle, placarding, maximum quantity carried, information carried and training of the driver.

637 Any fireworks must be carried and delivered to the site in appropriate packaging and in a form that has been classified by HSE. The details of these requirements are given on the CAD. The Classification and Labelling of Explosives Regulations 1983 and the Packaging of Explosives for Carriage Regulations 1991 control the classification and packaging of explosives.

Risk assessments

638 Clearly the planning and setting up of the display are important factors in ensuring the smooth and safe operation of the event.

639 Factors to consider when carrying out the risk assessment.

Display site location and layout

- Is the layout and size of the firing area adequate, considering the risk of the burning debris from one firework accidentally setting off another firework, and the need for firers to be able to move safely out of the area?
- Are the safety distances adequate for the fireworks to be fired, taking into account the risks from the malfunction of those fireworks, and other eventualities?

Setting up the fireworks

- Are the precautions to be taken while the fireworks are being set up adequate, taking into account risks to those doing the work and other people, including the general public?
- Have the risks from shells and other fireworks which can explode violently, or which project debris, been fully considered and adequate precautions taken?
- Have the noise levels from fireworks been taken into consideration? (Guidance on sound levels can be found in the chapter on *Sound: noise and vibration*).

Firing and clearing up

- Have the risks associated with these operations for the display been fully considered and adequate precautions taken?
- Where any event involves a work activity, the person who is providing the premises for the event may have duties under section 4 of the Health and Safety at Work etc Act 1974. It is likely that several employers will be involved at a music event. Management of Health and Safety at Work Regulations 1992 requires co-operation between the different employers to ensure that all risk-assessments are co-ordinated.

There is extensive guidance already available on the setting up and firing of fireworks. Consult HSE's publication (*Working together on firework displays*) for information on this issue.

Theatrical and stage pyrotechnics

640 Pyrotechnics are used in many productions from a small theatrical show to major rock concerts and as pyrotechnics are explosives, the dangers inherent in the types used for the entertainment industry should be understood.

641 All pyrotechnic effects produce light, colour, heat, sound and smoke or a combination of two or more of these elements. With one or two exceptions, pyrotechnics rely on the ignition of chemicals to create a combustion reaction; this can be spontaneous or over a longer period. Once ignited, pyrotechnic devices are virtually impossible to extinguish, therefore the effects must be chosen carefully.

642 To ensure safety:

- the effects should be obtained from a recognised manufacturer. Home-made effects are illegal and may be unreliable in performance;
- the effects should only be fired via a control system that has been designed and manufactured with adequate safety features built in, both mechanical and electrical;
- the user, ie the operator, should have enough experience and knowledge to ensure that, not only are the effects used correctly and safely, but that they can cope with any unforeseen circumstances.

643 One of the most common problems with the use of pyrotechnics is the lack of pre-planning. Pyrotechnics are often required but decisions are made very late with the result of trying to incorporate them into a show after everything else, eg the late siting of pyrotechnics on stage roofs. With early planning many of the problems can be overcome and unnecessary risks reduced.

Regulation

644 Theatrical and stage pyrotechnics are subject to the same legislative controls as those described for fireworks (see paragraphs 631-637).

Notification and inspection

645 Notify the local authority of the intended use of pyrotechnics in advance. The written notification should contain details including quantity, type, and a brief description of the effect type. Notification often includes a drawing or set plan showing the positions of each effect. If fireworks are going to be used near coastal water, the coast guard will need to be notified.

646 A risk assessment to cover the use of the pyrotechnics at the event should be prepared by a competent person. The insurance cover of the pyrotechnic company or the individual pyrotechnician should also be examined.

647 Arrange for an inspection of the proposed effects and their positioning with the operator. In a theatre this can often be done in advance of the show or opening night. With some events this is often impossible and the inspection normally takes place on the event day. The pyrotechnic content of any event is often looked at with other elements of the event, eg rigging, drapes, etc. It is often the last to go in - other than the cables, which are often the first - ie, the devices cannot be made ready until the set is in position.

648 A demonstration can be requested for any unfamiliar effects or if there are concerns as to the suitability in a certain position. This should take place where the effects will be positioned and fired during the event unless there is a very good reason for it not to be done *in situ*.

Safety considerations

649 Questions about any effect, firing system, operator, company, etc, should be resolved during the inspection.

650 Some aspects are as follows.

- Are the chosen effects safe in the chosen position? Is there enough safety distance to the set, the audience, other equipment and to any workers on stage?
- Which type of effects, if any, are at high level, for instance airbursts or waterfalls, and which type are close to the stage edge and vice versa? An example would be effects situated close to monitors, many of which, although built to withstand the rigours of touring, have foam linings which will burn quite easily.
- Some of the potentially most dangerous effects are maroon or concussion effects. Are there any, where are they positioned? Maroons or concussion must be situated away from all people and maroons can only be fired from a purpose-built bomb tank with the required warning notices in position. A warning light system is also a worthwhile additional safety precaution.
- Are any effects particularly noisy? Guidance on recommended sound levels for special effects are given in the chapter on *Sound: noise and vibration*.
- Check certificates for any drapes, especially if the event is outdoors where the drapes may have got wet.
- What state is the floor in? It should be solid, without cracks that could enable sparks to ignite stored items beneath it.
- Is any fall-out generated by the devices and if so where does this end up? Is this fall-out hot as well?
- Ensure no devices are situated over or across emergency exits from the venue.
- Where is excess material stored that is not to be used during the show? Some re-loading

may have to be carried out during the show although this is rare, or there may be several shows with pyrotechnics in the same complex. Excess material should be in a secure and suitable box and not left on the side of the stage.

- Check that adequate and suitable fire extinguishers are readily available (both water and carbon dioxide) and that workers know how to use them.

Firing and control systems

651 This aspect should form an essential part of the inspection. The controller must be one designed and manufactured specifically for pyrotechnics and not adapted from some other source. The most important feature of the controller is that it is operated via a removable key and that this is kept with the operator at all times and is put in the controller only for testing or firing. Under no circumstances should the controller key be in position at any other time.

652 The pyrotechnic technician must always have clear sight lines to the devices to be fired, to ensure there are no people close or that nothing has, for instance, been left on top of the devices. This is not always possible at larger events and often there will be two or more pyrotechnicians for larger events who keep in touch via radio or by CCTV.

653 All effects should be electrically fired, with one or two exceptions such as flash paper. Hand firing in an indoor situation is unnecessary and unsafe. Electrical firing means that, via a cable connected to an electrical igniter installed in a device, the effect is instantly fired and this has proved to be extremely reliable. Misfires/duds are very rare and if a device does not function immediately, there is virtually no chance that it will do so later. It is also fairly simple to physically disconnect any circuit that has a dud device on it therefore isolating it from any subsequent firings on the controller. Firing via transmitter/ receiver stations is now being used as well as computer based systems.

Operation

654 One person should be responsible for all aspects of the pyrotechnics, including the firing. The pyrotechnic technician has responsibilities to other workers, performers and audience, including the provision of appropriate personal protective equipment (the Personal Protective Equipment Regulations 1992). For instance, where there are noise effects, ear protection should be available. They should also ensure that a first-aid kit, eye wash kit and fire extinguishers are available.

Types of effects

655 Particular care should be taken with certain types that produce:

- sparks/fall-out which can remain hot for some time even after returning to ground level (eg star effects, gerbs, stage mines, saxons, airbursts);
- fall-out over a wide area (eg waterfall effect, saxons, airbursts);
- a considerable report (eg star effects, airbursts, maroons);
- considerable heat (eg coloured fire);
- a considerable amount of smoke (eg smoke puffs, coloured smoke, gerbs, waterfall effect);
- coloured dyes which can stain anything in the immediate area (eg coloured smoke);
- naked flame (eg flame effects);
- effects that can be fired or directed towards the audience (eg streamer/glitter/confetti units).

Cannons and maroons

Cannons

656 These devices must be used and positioned with care because of their power. They must never be pointed at people and should be firmly fixed, as there is considerable recoil when fired. The contents are propelled outwards with considerable force at great speed and can travel for up to 10 m before slowing down.

Maroons/concussion

657 Although one of the most popular and widely used effects, it is probably the most misused and potentially dangerous. Maroons explode and fragment with great force and produce a loud report. Concussion produces more of a 'boom' noise than a maroon. The concussion device is a heavy gauge small steel mortar, which needs the same safety procedures as maroons.

658 Maroons must only be used in a properly constructed bomb tank, not in dustbins, waste paper bins, old water tanks, etc. Bomb tanks should be sited off the stage area and well clear of all people and flammable materials. Careful positioning is essential and due care must also be taken with regard to light fittings or other items in the immediate vicinity of the tank. The best position for a bomb tank is sub stage. When bomb tanks are in use warning notices which read 'DANGER, EXPLOSIVES KEEP CLEAR, BOMB TANK IN OPERATION' should be positioned at all access points. Everyone must be kept away from the tank area. A warning-light system is also advisable and the relevant ear protection devices should be made available to all workers.

Chapter 18

Camping

659 At many events camping is no longer incidental and an alternative to other accommodation - it is an integral part of the event. The camping area should be provided within the defined event site and incorporated as part of the event planning. Emphasis needs to be placed on proper planning to ensure that an adequate level of services and facilities are available for the whole duration of the camping event and not merely during the licensed period of entertainment.

660 In isolated locations or where the music starts early or finishes late, contingency provision may have to be made for camping even when people were not intended to camp. Some consideration may also have to be given to crew camping and camping for stall holders with their stalls.

661 Services provided for people camping, including fire, stewarding, medical facilities, water supply, etc, need to be available for the length of time that campers are allowed to remain on the site. Ensure your event publicity states the opening and closing times of the campsite. If large numbers of campers are likely to remain after the event, consider a gradual closing of the site to encourage those people to move, but without exposing them to risk.

Site design

662 The camping area will need to be reasonably well drained and level with grass cut short to minimise the risk of fire spread. Camping should not be allowed on stubble. Break the camping areas up into discrete smaller areas to:

- provide an identifiable area for campers;

- allow for the management of each area;
- control the densities of each area;
- provide information and communications.

663 Music events involving camping are likely to attract a broad mix of people and it might be desirable to create a separate area for family camping. Separating areas can be carried out by simple measures such as posts and tapes while at larger events it may be necessary to provide some physical barrier to prevent camping such as metal trackways, ballasted roads, etc. Wherever possible the layout of the site should provide for an entertainment area in the middle of the site with camping on the periphery and parking beyond that. Crowd movements will therefore disperse away from the focus of the event. It is important that campsite layout plans are fully integrated between the various agencies involved, so that the site features and descriptions of locations will be identical for all the agencies.

664 Site arrangements and boundaries need to take account of natural hazards such as ponds, ditches, rivers, etc. Other hazards such as electricity pylons may need to be assessed to prevent access or risk of shock from activities such as kite flying and the use of tethered commercial balloons.

Site densities

665 Experience has shown that a density of up to 430 tents per hectare for rock/pop events is a realistic standard. At more family-orientated events, ie larger tents with greater number of occupants, this density would need to be reduced, possibly by around 50%.

666 It is desirable to provide separation distances between individual tents to make the site safer from fire and trip hazards, etc. Provide people entering the site with information and maps showing the camping areas and ensure there are sufficient stewards to direct people to the appropriate areas as the campsite fills up.

Segregation of vehicles/live-in vehicles

667 It is desirable to physically separate camping areas from vehicle parking areas. The reasons for this are to remove risks from:

- cruising or joyriding;
- car fires;
- runaway vehicles.

668 Try to minimise the distance between car parks and campsite. Consider providing internal transport for campers to and from the campsite. This is particularly important for families with children who need to carry considerable amounts of equipment.

669 It may be justifiable to permit parking with camping in certain circumstances on a level site and where the audience are compliant (eg families). Where there is a desire to allow camping and car parking next to each other, the density will need to be substantially reduced to allow for increased roads and separation. Particular attention will need to be paid to designing the campsite in advance so that blockages of tents and cars cannot happen. It may be acceptable to allow vehicles and tents to mix in an area provided for campers with special needs.

670 If live-in vehicles (eg dormobiles, camper vans or caravans or adapted vehicles) are to be allowed on site, set aside a special area for this purpose. Such vehicles should not be used for camping in a parking area.

Information, organisation and supervision

671 Include information on site restrictions, such as no unauthorised PAs, campfires, etc, on the ticket. At strategic points on the site (including the campsites) provide information including a 'You are here' map and key information to direct people to important facilities such as toilets, water, medical facilities, fire points, etc. Make information, including site safety and restrictions, easily available (in the case of a large event this could be by way of a dispersed warden service that would operate 24 hours a day). Ensure the warden service has radio communication and is able to respond to information requests from members of the audience about emergency situations involving individuals' health, fires, etc.

672 By breaking up the camping area into smaller discrete areas, people can be given an identifiable camping area to which they can more easily return. On complex sites involving many camping areas and a large entertainment area, provide all campers with maps on entry and/or preferably an information pack with safety advice.

673 Locate stewards within the camping areas before campers arrive to assist with the general build-up of the campsite, and to monitor key facilities such as toilets, fire provision, water supply, etc. These stewards will also have a role in helping to ensure that camping is dispersed in the best way over the designated camping areas.

Contingency planning

674 Aspects of contingency planning that require particular attention where there is camping on site include:

- adverse weather;
- failure of water supply;
- other need to clear the area.

675 At certain types of events attracting young people, it is not unusual for them to attend without tents. Similarly, people attending with tents may find that the tents are unusable so that they are without accommodation. Campers might also have their tents stolen. Therefore, contingency provision will have to be made to allow members of the audience to obtain shelter where they are unable to provide any themselves.

676 If temporary accommodation needs to be provided, existing marquees and tents may be suitable. In the case of adverse weather conditions, particularly wet weather combined with high winds, such structures may not be capable of being guaranteed stable. A source of smaller tents may therefore be advisable to provide emergency accommodation.

677 At large events where people arrive in large numbers by public transport it may be impossible to close the event and clear the camping area in an emergency. Facilities will have to be brought to the camping areas rather than the people removed to another place of safety.

Public health

678 It is useful to provide advice to individuals on basic personal hygiene matters and the sort of food

that they should or should not bring with them. Given the greenfield nature of a camping area, large numbers of people involved, basic sanitation and remoteness from care, it is essential to ensure that food outlets and personal hygiene are satisfactory. The consequences of an infectious disease outbreak would be significant in terms of both the numbers that could be involved and the likely amount of care that could be provided. Provide adequate catering facilities, some overnight, and outlets where campers can buy basic provisions such as bread, vegetables, milk, etc.

679 Sites that are grazed will naturally be contaminated with animal droppings and may expose campers to health risks such as *E. coli* 0157 infection. Exclude animals from all areas other than car parks for as long as possible before public access. *E. coli* 0157 can survive for long periods in the environment.

680 Dogs should not be permitted on the site and advance publicity should be given. Unnecessary health risks include fouling and dog bites, and stray dogs pose a nuisance. Notwithstanding any advance publicity, it is likely that people will bring dogs, in which case provision should be made to deal with strays.

Crime

681 Campers are vulnerable to having property stolen from tents but may be unable to carry around items that might be stolen if left unprotected in their tents. Consider providing secure accommodation on campsites where people can leave bulky or valuable items.

682 Campsites should be adequately lit and patrolled by stewards to deter both isolated and organised criminal activity. Patrols will also help to identify other matters such as fire outbreaks, camp fires getting out of control, etc.

Fire safety

683 Campfires constitute a risk of burns, tent fires and can cause smoke pollution. They are undesirable and wherever possible should be discouraged. At some types of event however, it would be virtually impossible to prohibit fires and for certain audience profiles more regulated (communal) fires are unlikely to be an attractive option. Where fires are allowed, consider the material campers will burn. Consider providing chopped firewood to avoid destruction of trees and hedges and the potential for burning plastics and other material that could produce noxious fumes.

684 Consider the hazards and risks of camp fires in the event risk assessment to include the following:

- suitably trained stewards or fire marshals;
- fire points: as a minimum these should consist of a means of raising the alarm such as a gong or triangle and supplies of water and buckets, although these are probably of limited use in a tent fire;

- watchtowers consisting of raised platforms staffed by stewards with radios are a more effective means of observing for uncontrolled fires and suspicious behaviour. They should be supplemented by the provision of fire extinguishers and, depending on the scale of the event, an on-site capability to attend to fires with specialised vehicles;
- the fire points themselves becoming a hazard due to rubbish accumulation, etc.

Site services

685 Ensure that facilities are maintained throughout the site 24 hours a day and services are provided for the duration that people are actually on site. All facilities must be lit at night.

First aid

686 See the chapter on *Medical, ambulance and first-aid management*. At camping events that run over several days it will not be sufficient to provide a first-aid facility only. Expect the demands that would be placed on a GP practice serving a community of similar size. Routine medical supplies, therapeutic drugs, etc, may need to be provided, including pharmacy facilities, dentistry and psychiatric facilities.

Welfare

687 See the chapter on *Information and welfare*. In addition, it is likely that there will be a number of children present on site and facilities will have to be provided, including potentially accommodating children overnight. Communications and availability of information on lost children, lost friends, etc, must be established (see chapter on *Children*).

Telephones

688 Provide telephones in suitable numbers and ensure they are easily accessible, well signed and available 24 hours a day to cater for the demand from individuals to contact parents, friends, etc.

Sanitary facilities

689 See the chapter on *Sanitary facilities*. In the case of events with large camping areas, it is not sufficient to merely allocate facilities on a numeric basis; assess where and when facilities will be under pressure. There will inevitably be a peak morning demand.

690 It is suggested that a plan is established whereby sanitary accommodation, drinking water supplies, washing facilities and showers are all clustered together, creating an easily identifiable location for all facilities. Monitor the condition of sanitary accommodation to ensure they are regularly emptied and cleaned as required in addition to routine programmed servicing.

Refuse

691 Provide refuse receptacles along the walkways and access ways for vehicles and also at conspicuous points such as sanitary facilities, etc. Ensure that bins are emptied on a regular basis to encourage careful disposal and to avoid creating a fire hazard. On greenfield sites with potentially difficult terrain this is likely to be achieved by tractors and trailers. Reductions in volume of refuse are likely to be achieved by using recycling points to take separated waste.

Site lighting

692 Provide adequate lighting to enable orientation at night, with higher levels of lighting at toilet areas, fire points, information and warden points, etc. Take care in deciding on the nature of lighting. Lighting tower rigs are likely to be unsuitable for camping areas due to generator noise as well as providing an overly bright source of light. They may, however, be suitable for junctions, crossroads and facilities, etc. Festoon lighting can be tampered with so it won't work or becomes a safety hazard. Wherever possible, provide the camping areas with some illumination provided from 'borrowed light' from other areas of higher lighting nearby, which can be supervised.

Access

693 Provide both vehicular and pedestrian tracks to and through camping areas to ensure ready access for emergency vehicles and also to provide safe routes for pedestrians free of trip hazards such as guy ropes, etc.

Noise

694 Plan for preventing or reducing the impact of potentially noisy activities within campsites or of dealing with any overnight activities that become problematic. Dependent upon the nature and proximity of residences to the site, restrictions may be needed in limiting the background music provided by concessionaires to avoid noise disturbances.

Chapter 19

Facilities for people with special needs

695 Consider suitable arrangements, wherever possible, to ensure that all people with special needs are able to attend. It is also recommended that a complete access strategy is prepared which includes the technical issues as well as factors which will encourage and attract persons with special needs to your event. Details of national organisations who may be able to provide guidance are given at the end of this book, in *Useful addresses*.

696 Consider provision for people with:

- mobility problems (including wheelchair users);
- difficulty in walking;
- impaired vision and/or hearing.

697 Event publicity should provide a contact number where people with special needs can obtain information on site arrangements.

698 When designing your site or venue consider how people with special needs can best be accommodated. This includes easy access and adequate means of escape in an emergency. The number of wheelchair users who can be admitted will be dependent upon a number of factors including the structural and internal layout of the venue.

699 Wheelchair spaces in parts of a seated area should allow for adequate room for manoeuvring a wheelchair. Generally, a manual wheelchair needs approximately 0.9 m width and 1.4 m depth. Electric wheelchairs need more space.

Access

700 Place parking facilities for people with special needs at the most directly accessible point to those areas set apart for wheelchair users. Spaces allocated should be wider than normal (about 3.6 m) to allow room to manoeuvre. At outdoor events parking for people with special needs should also be placed at the most directly accessible point to the allocated seating areas, as well as the most directly accessible point to designated and accessible campsites. Thought should be given to the means of having direct and safe access links between the designated parking, camping and seating areas. Use flat surfaces or ramps to provide access from parking or drop-off areas to designated areas.

Ramps

701 Ramps for wheelchairs should conform to BS 5810: 1979. The ramp should have an easy gradient and it is suggested that it should not be steeper than 1 in 12. Ramps should have a level resting space landing every 10 m. They should also have raised safety edges and handrails.

Viewing areas

702 As standing audiences can cause surging movements, all people attending the event that have any mobility difficulties should be located in an area where they will not be affected. When setting aside viewing areas for people with special needs, the area should have a clear view of the stage, often beside the 'mixing' tower. The area should be constructed using non-slip materials with direct access to an exit.

703 At outdoor concerts wheelchair users can be accommodated either on an open area or on a flat terrace with direct access to toilet facilities and concessions. The eye level of a wheelchair user is estimated as being between 1.1 m and 1.25 m.

704 Many wheelchair users will be accompanied by an able-bodied companion. Make sure that space in the wheelchair users area can accommodate these companions, preferably with chairs provided which do not block the view of other wheelchair users in the area.

Facilities

705 Concession stands should also be encouraged to have either varied level of serving counter space or an access ramp in front of the serving counter. Toilet allocation (see chapter on *Sanitary facilities*) should be unisex cabins with wheelchair access and it is suggested that one unisex cabin per 75 wheelchair users should be provided, along with additional provision for the use of carers, etc.

Support

706 Stewards or special needs assistants should be in attendance to ensure that facilities which are provided for people with special needs are available for the intended purpose.

707 Consider providing designated 'ground support' workers. They could be people with special skills (signers, medics, etc) who can provide on-site support for people with special needs. These workers should be easily recognisable by the use of an easy-to-read emblem or logo, eg the letter 'A' for 'assist' emblazoned on an outer garment. Stewards operating in and near to the area set aside for people with special needs require training in the evacuation and exit procedures. Also consider using safe sites for people with special needs in the event of an evacuation.

People with impaired vision

708 People with impaired vision or colour perception may have difficulty in recognising information signs including those used for fire safety. Signs therefore need to be designed and positioned so that they can easily be seen and are distinguishable. Good lighting and the simple use of colour contrasts can also help visually impaired people find their way around. Where practicable, consider admitting guide dogs. Advice on any of these matters can be obtained from the Royal National Institute for the Blind or the National Federation of the Blind of the United Kingdom.

Evacuation

709 People in the audience may be affected by a range of disabilities, including restricted mobility, epilepsy, impaired hearing, mental health problems, etc, so their needs and requirements should be included in major incident and contingency plans. Where they exist, electronic display systems should be used to give information, including evacuation messages, particularly for people with impaired hearing.

Publicity

710 It would be helpful to potential visitors if the facilities that are available are publicised. This can be achieved by contacting the local disability association, access groups and local clubs or organisations for people with disabilities.

Further information

711 Organisations offering support and advice for people with special needs can be found in *Useful addresses*.

Chapter 20

Medical, ambulance and first-aid management

712 The aim of this chapter is to set out the responsibilities of the event organiser to ensure that medical, ambulance and first-aid assistance, as appropriate, are available to all those involved in an event. The event organiser needs to minimise the effects of an event on the healthcare provision for the local population and, wherever possible, reduce its impact on the local NHS facilities and ambulance service.

713 The number of people requiring medical treatment at any music event will vary considerably as will the type of ailment. These will vary with environmental conditions and can range from traumatic injuries due to crushing, falls, fighting or conditions such as hyperventilation, exhaustion, dehydration, sunstroke, hyperthermia or hypothermia, emotional or anxiety attacks, food poisoning or the serious effects of drugs or alcohol. Acute medical emergencies such as heart attack or stroke will need to be provided for as well.

714 At events which may take place over several days, such as festivals, conditions common in general practice are likely to predominate. In addition, people with various existing disabilities and medical conditions such as asthma, diabetes, heart, or psychiatric problems may attend events where their condition could be worsened.

715 Previous experience suggests that approximately 1-2% of an audience will seek medical assistance during an event day. Of these, around 10% will need further treatment on site. Approximately 1% of the number requiring initial medical assistance will require subsequent referral to hospital. It should be recognised that other factors such as ineffective welfare facilities, poor weather conditions, absence of free drinking water or the presence of other 'on site' hazards may increase this number.

716 It is essential that all major music events have suitable arrangements for the triage, treatment, and transport of those in need. Ensure that this provision is approved by the local authority who will take into account the recommendations of the NHS usually through the NHS ambulance service or in Scotland the local health board.

Planning

717 Plan the provision of medical, ambulance and first-aid services along with the statutory services and appoint a competent organisation to provide medical management. This organisation need not be the sole provider of resources at the event, but must be able to demonstrate competence in operating the medical arrangements. In addition, the appointed organisation should be experienced in the medical management of similar events, and must accept responsibility for providing an appropriate management and operational control infrastructure and co-ordinate the activity of other medical providers. Ensure that the appointed medical provider liaises with other statutory services and first-aid providers on site. Respective roles and responsibilities should be set out in a medical, ambulance and first-aid plan.

718 It is considered good practice to consult with the local NHS health authority ambulance service, (local health board in Scotland) for the area so that they can advise both the event organiser and the local authority on the likely impact of the event on pre-hospital accident and emergency services in the area.

Named manager

719 A manager from the medical provider should be appointed to take overall control and co-ordination of first-aid provision. This person should also be readily available during the event. The event organiser and the appointed medical provider should liaise with all interested parties which may include the local NHS health authority, health board, ambulance service or competent first-aid providers, as appropriate.

Confirmation

720 It is recommended that the final details of the event are confirmed in writing to the appointed medical provider as soon as possible.

Specific considerations

Build-up and breakdown

721 Consider the availability of medical, ambulance, and first-aid provision during the build-up and breakdown of the event (see *First aid for employees and event workers*, paragraph 757-758).

Queuing

722 Consider the need for medical, ambulance and first-aid arrangements for any audience members queuing before the gates or doors open and when they leave at the end of the event.

Information

723 Information on the location of first-aid facilities must be available to all those attending. Provide adequate signage and consider printing the location of first-aid facilities on tickets for the event. In addition, stewards should be aware of the nearest facility.

Campsites

724 At events with overnight campsites, appropriate provision should be made to have medical, ambulance, and first-aid cover available while the campsite is open. Because of the likely range of conditions requiring medical advice, also consider GP services through the appointed medical provider during the times the campsites are in operation.

Sterile routes

725 Where practicable, consider the provision of suitable sterile routes for the exclusive use of emergency vehicles.

Location

726 The location of responders is important when assessing the response times for the arrival of emergency care to individual casualties at any location within the event.

Vehicle movement

727 Only in exceptional circumstances should ambulance vehicles be allowed to enter audience areas. Ambulances should not move from their designated position except on the instruction of their control unless compromised on grounds of safety. At events with high audience densities consider the use of foot squads or buggies to remove casualties.

Maintaining cover

728 The appointed medical provider should have in place arrangements to ensure that cover is maintained at the correct level throughout the event. If a casualty needs to be removed from the site by ambulance, arrangements must be in place to replace that vehicle or to transport the casualty using an ambulance dedicated to off-site patient movement (if there is the need for ambulances on site).

Helicopters

729 At certain events, an area for medical evacuation by helicopter may be required and a suitable landing site, either at the site or nearby, identified.

Communications

730 At large events, there may be a need for a separate medical radio channel connecting the NHS ambulance service with ambulance workers, key medical workers, mobile response teams, and key first aiders. A protocol for the use of radio equipment, including consistent call signs, must be agreed before the event. A communications plan detailing medical communications links should be produced and held at both the medical control point or incident control room and central ambulance control.

731 If there is more than one medical facility, there should be a designated main medical facility with an external telephone line (which does not go through a switchboard) and a list of appropriate numbers. All other medical facilities should have an internal telephone or radio link to the main position.

Documentation

732 An event log should be maintained, which should include any actions or decisions taken by the manager of the medical provisions and the reasons for those actions.

Note: Event logs, report forms and records completed at an event may be required at a later date to assist in the reporting of accidents and injury to workers and audience members under the Reporting of Injuries, Diseases and Dangerous Occurrences Regulations 1995 (RIDDOR).

733 Ensure that the appointed medical provider maintains a record of all people seeking treatment. In some locations, for consistency and ease of documentation, suitable patient report forms may be supplied by the NHS ambulance service. This record should include details such as: name, address, age, gender, presenting complaint, diagnosis, treatment given, the onward destination of casualties (eg home, hospital, own GP), and the signature of person responsible for treatment. The only people who may be shown patients' records are those that are involved in the treatment or those that have lawful authority.

Medical, ambulance and first-aid provision

734 Following the risk assessment and agreement on levels of medical, ambulance and first-aid cover, this and any further arrangements relating to health care should be written in a statement of intent and signed by the relevant parties and included in the event management plan. Ensure that a suitable skills mix exists and that medical, ambulance and first-aid providers are located effectively throughout the site.

735 The decision on the level of medical provision and whether the NHS ambulance service will be directly involved, or not, at any particular event will depend on a number of specific factors including:

- size of audience;
- nature and type of event and entertainment;
- nature and type of audience - including age range;
- location and type of venue - outdoor or indoor, standing or seated, overnight camping and the size of the site;
- duration of event - hours or days;
- seasonal/weather factors;
- additional activities and attractions;
- proximity/capability/capacity of local medical facilities;
- intelligence from other agencies regarding previous experience of similar events;
- availability and potential misuse of alcohol or drugs (illicit, recreational, or controlled);
- external factors including the complexity of travel arrangements;
- time spent in queues;
- availability of facilities on site including welfare, befriending and other social services;
- range of possible major incident hazards at or associated with the event (structure collapse, civil disorder, crushing, explosion, fire, chemical release, food poisoning);
- availability of experienced first aiders.

736 Tables are provided at the end of this chapter, pages 131-134, which show a way of calculating the quantities of medical, first aid and ambulance provision required for various event types.

First aiders

737 The recommended minimum number of first aiders at small events where no special risks are considered likely is 2:1000 for the first 3000 attending. No event should have less than two first aiders.

738 At indoor venues or stadia, first-aid facilities are likely to have been agreed. However, the historical number of first aiders provided at an existing venue does not replace the need to carry out an assessment for each event. Some venues will be in multiple use. In such cases, the overall provision of medical, ambulance and first-aid resources should take account of all activities taking place within that venue.

739 At events attended by a very young audience, and at long events or street carnivals, the number of first aiders may need to be significantly increased or the ratio of first aiders to professional ambulance workers, doctors or nurses altered. In these circumstances, the ambulance provision required should be discussed with the local NHS ambulance service, who may recommend special requirements. This may include the provision of an ambulance control unit or an ambulance emergency equipment vehicle.

Medical practitioners

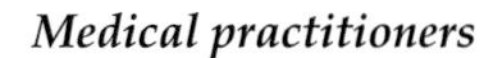

740 The risk assessment may indicate the need for the provision of medical practitioners on site. Any medical practitioners should be provided in addition to any medical workers allocated for the care of performers.

741 Usually one suitably experienced medical practitioner should be able to fulfil the role of medical incident officer with overall responsibility, in close liaison with the ambulance incident officer for the management of medical resources at the scene of a major incident.

Psychiatric care

742 At lengthy or large events, consideration should be given to any requirement for a psychiatric care team including psychiatrists, psychiatric nurses and drug advisers. This team may need to liaise with the local authority social services department, hospital authorities and the police.

Nurses

743 Qualified nurses may be required to care for patients requiring longer-term management on site. Unless trained as part of a mobile-response team, nurses should undertake the specific duty of staffing the main medical facility, working as a team with the medical practitioners, paramedics and first aiders in the triage and treatment of casualties.

Paramedics, ambulance technicians and ambulance care assistants

744 Paramedics and ambulance technicians may need to be positioned in the pit area, medical facilities or areas of perceived risk, or deployed in immediate response to emergencies arising throughout the event area.

745 Ambulance care assistants may assist in the transport of those with non-urgent medical conditions or with minor injury.

Medical cover in pit areas

746 The risk assessment may indicate that medical cover may be required within the pit area. Medical workers in this area should be suitably experienced and trained to provide advice on casualty handling to stewards, appropriate triage to casualties and, where required, have the ability to facilitate the rapid evacuation of any casualties to a medical facility. The area in front of the stage should have the following equipment quickly available:

- rescue board and cervical collars;
- oxygen therapy and resuscitation equipment;
- assorted splintage.

On-site medical facilities (first-aid points)

747 The number, location and suitability of medical facilities should be planned. If there is more than one medical facility, one should be designated as the main medical facility. First-line medical facilities, including those in the pit area, will refer those requiring further treatment to the second-line main medical facility. The main medical facility may be equipped as a medical centre or field hospital. In the event of a major incident, in accordance with local major incident procedures, a medical facility will be established or designated as the casualty clearing station.

Maps and plans

748 Detailed gridded maps or plans of the site with position of medical facilities clearly marked must be available before the event. This should include the surrounding roads and access routes.

Structures

749 At outdoor events, if a suitable permanent structure is not available, provide suitably equipped mobile first-aid units or marquees with appropriate flooring. At indoor events, position the medical facility in or next to the main arena.

Staffing plan

750 An appropriate number of competent first aiders should staff each medical facility and, as appropriate, medical workers, ambulance workers and nurses, some of whom should be available to offer assistance within audience areas. At large outdoor events, ensure that a proportion of mobile first aiders are strategically positioned or asked to patrol a defined area, in consultation with the NHS ambulance service, if present. All workers must be clearly identified. Mobile first aiders should be in constant radio contact with their controller.

Mobile response teams

751 At high-risk events, consider the use of a suitably equipped mobile response team with an appropriate skills mix and means of transport to attend medical emergencies where their specific skills are required.

Position

752 At larger events, provide a medical facility near to the stage area with unrestricted access to this position from the pit area. In general, other medical facilities are situated in positions on the perimeter of the audience area enabling unrestricted access and exit for ambulances without entering areas occupied by the audience.

General considerations for the main medical facility

753 As a minimum requirement, the main medical facility should be:

- designated as a 'no smoking area';
- of an adequate size for the anticipated number of casualties and readily accessible for the admission of casualties and ambulance crews;
- large enough to contain at least two examination couches or ambulance stretcher trolleys, with adequate space to walk around, and an area for the treatment of sitting casualties;
- accessible at ground level and have a doorway large enough to allow access for an ambulance stretcher trolley or wheelchair;
- maintained in a clean and hygienic condition, free from dust and with adequate heating, lighting and ventilation;
- provided with adequate first-aid and medical equipment and screens, etc, including resuscitation equipment, patient-care consumables and where appropriate, a defibrillator, all of which should be separate from those contained in ambulances. An agreement should be reached during the planning stage about who will provide such items;
- within close proximity of an easily accessible wheelchair-users toilet and workers facility;
- provided with a supply of running hot and cold water. If this is not possible, provide adequate fresh clean water in containers;
- provided with a supply of drinking water over a sink or hand-wash basin or suitable receptacle;
- provided with a worktop or other suitable surface for equipment and documentation, eg folding tables;
- provided with suitable secure storage facilities for drugs and equipment used by the medical providers;
- next to appropriate hard standing or parking facilities for ambulances or associated emergency vehicles.

Clinical waste

754 Specific arrangements for the disposal of clinical waste must be planned. Special 'Bio hazard' containers for the disposal of 'sharps' or appropriately marked 'yellow bags' for the disposal of dressings or other contaminated materials will be required. Suitable arrangements must also exist for the disposal of non-clinical waste at medical facilities.

Liaison with welfare services

755 The workers at medical facilities should be made aware of the arrangements for social/welfare provision so that people can be suitably redirected to those facilities.

Welfare of employees and event workers

756 Plan the welfare of the medical, ambulance, nursing and first-aid workers. At any event which lasts more than four hours, provide rest areas, sanitary and dining facilities. Where possible, separate these areas from the audience facilities.

First aid for employees and event workers

757 Under the Health and Safety (First Aid) Regulations 1981, employers are responsible for ensuring that first-aid facilities, equipment and personnel are provided for their employees if they are injured or become ill at work. In order to decide on the level of first-aid provision necessary, an employer should make an assessment of the first-aid needs appropriate to the circumstances of the workplace. Employees who are appointed as first aiders must have successfully completed the necessary training with an HSE approved training organisation. It is also good practice to have an accident book available in which to record incidents which require first-aid treatment. It is strongly recommended to have a written agreement between the various employers, eg contractors, subcontractors and others working at the event to ensure that the first aid provided meets all their needs and to avoid misunderstandings.

758 Further guidance on Health and Safety (First Aid) Regulations is contained in HSE's publication *First aid at work: The Health and Safety (First-Aid) Regulations 1981 Approved Code of Practice and guidance.*

Definitions and competencies for medical workers

759 First aiders, ambulance and medical workers should:

- be at least 16 years old and not over 65 years old;
- have no other duties or responsibilities;
- have identification;
- have protective clothing;
- have relevant experience or knowledge of requirements for first aid at major public duties;
- be physically and psychologically equipped to carry out the assigned roles.

Also, first aiders under 18 years old must not work unsupervised.

Medical practitioner

760 A 'qualified medical practitioner' is a medical practitioner registered with the General Medical Council in the UK. The medical practitioner should be familiar with, or have access to, the local authority and NHS major incident plans and have completed a course in major incident management. The medical practitioner should also have recent experience in dealing with emergencies in the pre-hospital or accident and emergency environment (within two years) and be familiar with the operation of the local NHS ambulance service and competent first-aid providers. In addition, he or she should have attended a course in pre-hospital emergency care.

Qualified nurse

761 A 'qualified nurse' is a nurse whose name is entered in the relevant part of the professional register maintained by the UK Central Council for Nursing, Midwifery and Health Visiting. The qualified nurse should have post-registration knowledge and recent experience in dealing with emergencies in the pre-hospital or accident and emergency environment (within two years).

Paramedic

762 A 'paramedic' is a member of an NHS ambulance service who holds a current certificate of proficiency in paramedical skills, issued by the Institute of Health Care and Development (IHCD), and who has immediate access to the appropriate level of specialist equipment, including drug therapy, as stipulated and approved by the relevant Paramedical Steering Committee.

Ambulance technician

763 An 'ambulance technician' is a member of an NHS ambulance service who holds a current certificate of proficiency in ambulance aid skills issued by the IHCD.

Ambulance care assistant

764 An 'ambulance care assistant' is a member of an NHS ambulance service who has completed an ambulance care assistant course at an IHCD recognised ambulance training establishment.

First aider

765 A 'first aider' is a person who holds a current certificate of first-aid competency issued by any of the three voluntary aid societies (or certain other bodies or organisations): St John's Ambulance, British Red Cross Society or St Andrew's Ambulance Association. The first aider should have prior training or experience in providing first aid at crowd events.

766 **Note**: The completion of a 'Health and Safety at Work' or four day 'First Aid at Work' course does not necessarily qualify a person as competent to administer first aid to members of the public.

Appointed medical provider

767 A competent organisation chosen by the event organiser, to provide overall management of medical, ambulance and first-aid services at an event.

Medical, ambulance and first-aid provision

768 It is recognised that medical cover at events can be organised in different ways and that the most appropriate model will vary according to the medical provider and the nature of the event. The following tables set out a method of estimating a reasonable level of resource.

769 It is emphasised that these figures may require modification as some providers may choose to substitute medical staff or paramedics for first aiders. In any case, the suggested levels of resource are intended only as general guidance and should not be regarded as prescriptive. The tables are not a substitute for a full risk assessment of the event. Figures do not take account of dedicated cover for performers or VIPs.

- Use Table 1 to allocate a score based on the nature of the event.
- Use Table 2 to allocate a score based on available history and pre-event intelligence.
- Use Table 3 to take into consideration additional elements, which may have an effect on the likelihood of risk.
- Use Table 4 to indicate a suggested resource requirement.

Table 1 Event nature

Item	Details	Score
(A) Nature of event	Classical performance	2
	Public exhibition	3
	Pop/rock concert	5
	Dance event	8
	Agricultural/country show	2
	Marine	3
	Motorcycle display	3
	Aviation	3
	Motor sport	4
	State occasions	2
	VIP visits/summit	3
	Music festival	3
	Bonfire/pyrotechnic display	4
	New Year celebrations	7
	Demonstrations/marches/political events	
	Low risk of disorder	2
	Medium risk of disorder	5
	High risk of disorder	7
	Opposing factions involved	9
(B) Venue	Indoor	1
	Stadium	2
	Outdoor in confined location, eg park.	2
	Other outdoor, eg festival	3
	Widespread public location in streets	4
	Temporary outdoor structures	4
	Includes overnight camping	5
(C) Standing/seated	Seated	1
	Mixed	2
	Standing	3
(D) Audience profile	Full mix, in family groups	2
	Full mix, not in family groups	3
	Predominately young adults	3
	Predominately children and teenagers	4
	Predominately elderly	4
	Full mix, rival factions	5
Add A+B+C+D	***Total score for Table 1***	

Table 2 Event intelligence

Item	Details	Score
(E) Past history	Good data, low casualty rate previously (less than 1%)	-1
	Good data, medium casualty rate previously (1% - 2%)	1
	Good data, high casualty rate previously (more than 2%)	2
	First event, no data	3
(F) Expected numbers	< 1000	1
	< 3000	2
	< 5000	8
	< 10 000	12
	< 20 000	16
	< 30 000	20
	< 40 000	24
	< 60 000	28
	< 80 000	34
	< 100 000	42
	< 200 000	50
	< 300 000	58
Add E+F	***Total score for Table 2***	

Note: Numbers attending may vary throughout the duration of longer events. Therefore, resource requirements may need to be adjusted accordingly.

Table 3 Sample of additional considerations

Item	Details	Score
(G) Expected queuing	Less than 4 hours More than 4 hours More than 12 hours	1 2 3
(H) Time of year (outdoor events)	Summer Autumn Winter Spring	2 1 2 1
(I)Proximity to definitive care (nearest suitable A&E facility)	Less than 30 min by road More than 30 min by road	0 2
(J) Profile of definitive care	Choice of A&E departments Large A&E department Small A&E department	1 2 3
(K) Additional hazards	Carnival Helicopters Motor sport Parachute display Street theatre	1 1 1 1 1
(L) Additional on-site facilities	Suturing X-ray Minor surgery Plastering Psychiatric / GP facilities	-2 -2 -2 -2 -2
Add G+H+I+J+K Subtract L	***Total score for Table 3***	

Calculation

To calculate the overall score for the event, do the following:

Add the total scores for Tables 1+2+3 above to give an overall score for the event

Table 4 Suggested resource requirement

Use the score from the above calculation to gauge the levels of resource indicated for the event.

Note: The following is an indication of the resources that may be required to manage an event based an assessment of factors set out in the previous tables. It must be noted that this table, in conjunction with the medical chapter, is intended for guidance only. It cannot encompass all situations and is not intended to be prescriptive.

The score refers to the suggested resources that should be available on duty at any one time during the event and not the cumulative number of personnel deployed throughout the duration of the event.

Score	Ambulance	First aider	Ambulance personnel	Doctor	Nurse	NHS ambulance manager	Support unit
<20	0	4	0	0	0	0	0
21-25	1	6	2	0	0	visit	0
26-30	1	8	2	0	0	visit	0
31-35	2	12	8	1	2	1	0
36-40	3	20	10	2	4	1	0
41-50	4	40	12	3	6	2	1
51-60	4	60	12	4	8	2	1
61-65	5	80	14	5	10	3	1
66-70	6	100	16	6	12	4	2
71-75	10	150	24	9	18	6	3
>75	15+	200+	35+	12+	24+	8+	3

Note: An ambulance paramedic crew, as a minimum, consists of a paramedic plus an ambulance technician trained to IHCD standards.

Chapter 21

Information and welfare

770 Providing information and welfare services at an event not only contributes to the safety and well-being of the audience but they also act as an early-warning system to detect any potential breakdown of services or facilities on site. The range and level of information and welfare services needed at any event will be determined by the event risk assessment.

771 Make sure that you clarify the role and responsibilities of welfare and information workers in advance. Fully brief other services involved in the organisation and management of an event, such as stewards and emergency services about the nature of available welfare and information services and encourage them to share information and liaise with such services before and during an event. Ensure that workers in the information and welfare services have suitable access and communication with members of the event management team, stewards, first aid, etc.

772 Locate information and welfare services in suitable accommodation, easily accessible, well signed, properly lit and make sure that they are open for the whole time the audience are on site.

Information

773 Information is an essential element in crowd management. Research has shown that when people have difficulty in obtaining information, they may feel unsatisfied, discontented, or even become aggressive. In turn, this may result in people becoming less likely to comply with safety instructions or in the extreme, lead to public order problems.

774 Provide advance information about the site layout and facilities at the point of sale of tickets,

preferably in the form of a leaflet. This could include welfare provision, crime prevention advice, suggestions for clothing, food and shelter, personal security, essential health and safety measures, a site map, meeting-up arrangements, transport and parking details, information about any prohibited items or practices, and details about how changes in event information (eg line-up, transport, location of services) will be relayed during the event.

775 Consider using the ticket itself to show the site in relation to main routes, dates of the event and any safety information or special conditions. This could be backed up by providing a telephone number for further information on the entertainment, and facilities available for people with special needs.

776 Provide information points according to the scale and duration of the event, to cover all times the audience are on site. Locate them in prominent places, well displayed and lit. Site maps should be freely available or at least provided for reference.

777 Information points can give details of public transport, on-site facilities, local facilities, performance times, message-leaving systems and access to emergency services by telephone. Some of this information could also be provided, along with other safety advice, on free handouts or in the event programme. The information point can also be used as a resource to provide up-to-date safety messages. An external telephone line or radio for emergency or humanitarian use is advisable.

778 Brief the workers in the information and welfare services with the site services and layout, as well as emergency procedures and facilities for dealing with lost or distressed people.

779 Display the site plan prominently at entrances, information points, car parks, first-aid points and in the event programme. Ensure that plans are large, clear, preferably in large print, waterproof and show the following information, as relevant:

- toilets
- performance areas
- camping areas
- exits and entrances
- car parks
- main roads
- first-aid points
- emergency services
- fire points
- welfare points
- information points
- police point
- catering facilities
- lost people's meeting point
- lost children's facility

- public telephones
- children's play areas
- lost property
- drinking water
- emergency shelter(s), where applicable
- property lock-ups, where applicable.

780 The issue of pass outs will need to be addressed as part of the event risk assessment. It is important that the audience is warned in advance if pass outs to the car parks or the local area are not allowed.

Welfare services

781 Welfare services are provided for people who find themselves in difficulties. These services fill in gaps not provided by other specialist services such as medical services, police and stewards. They need to be open during the whole time the audience are on site. Ensure that welfare workers are competent and have received adequate training and briefing.

782 Workers from day centres, night shelters, drugs projects, counselling services, alcohol services, probation, social services, mental health workers, teachers and solicitors, all offer relevant experience which could be invaluable in a welfare service setting. Welfare services, however, offer crisis-intervention and cannot offer long-term support.

783 Consider incorporating welfare services into your major incident and contingency planning for an event. Welfare services are well placed to offer a wide range of support to witnesses of incidents or to relatives of those involved in incidents. Welfare and information services can help with the dissemination of information during and after emergencies and incidents.

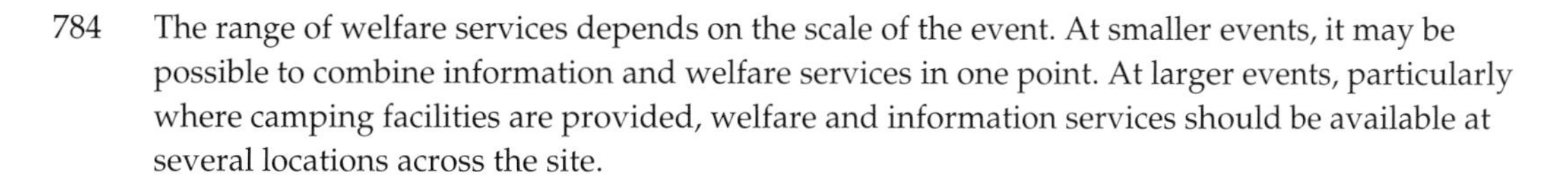

784 The range of welfare services depends on the scale of the event. At smaller events, it may be possible to combine information and welfare services in one point. At larger events, particularly where camping facilities are provided, welfare and information services should be available at several locations across the site.

785 Given the use of drugs and alcohol by young people, it is essential that you consider the provision of appropriate drugs and alcohol counselling, if relevant.

Meeting point and personal messages

786 One of the most common problems at events is that people become separated from the group with whom they have travelled. When planning your venue or site design consider a place for people to leave messages for each other and a clearly marked meeting place for people to meet up. A card index system is found to be the most effective way of leaving messages.

787 There may also need to be a meeting point just outside the site for meeting up or collection by relatives after the event. It is especially important that the meeting point is well lit and clearly signposted and does not create an obstruction to exits. The use of the PA system to locate individuals in the audience in consultation with the welfare services workers is recommended.

Public telephones and other public services

788 Consider providing public telephones especially if the site is isolated and if pass outs or re-admissions are restricted. A choice of coin and card payphones is best. For security reasons and to encourage 24-hour service, cardphones are preferable to cashphones, provided that cards are available for purchase on site. It is essential that there is an operator connection to enable reverse charge calls. Public telephones need to be accessible and operate during an emergency. Wherever possible, encourage banks to provide a service on site.

Lost property point

789 Information and welfare services can also provide a lost property point to deal with property found on site and where missing or stolen property can be reported (where appropriate, in liaison with police services on site).

790 Locating a lost property point with welfare services has the additional advantage of providing support for people who become distressed because of losing their possessions. However, a separate entrance for the lost property point may be advisable at larger events where the amount of lost property enquiries could overwhelm regular welfare and information services.

791 Make sure that found property is taken to the lost property point where details of the property, where it was found, and who found it are recorded. You may also need to make arrangements for the safekeeping of property which has not been reclaimed at the end of an event.

Cloakroom/left luggage/lock-ups

792 Providing lock-ups on site where the audience can leave valuable or bulky items may reduce the amount of lost property and theft. The type of items to be deposited and the design of the provision will depend on the type of event, eg camping equipment, rucksacks, etc, at camping events; and coats, bags and warm clothing at one-day events.

Support for vulnerable attendees

793 Provide support for members of the audience who become distressed during the event or who are deemed vulnerable. Such services are often also able to provide a space where disorientated people can feel safe and can stay until they feel able to leave, or need referral to specialist help. At larger or longer events, there may also be a need to arrange access to specialist services for people in acute need of psychiatric assessment and support.

794 There may also be a need to have competent workers available to help attendees experiencing difficulties through drugs they have consumed, including alcohol. Basic guidance on assisting people with drug-related difficulties can be obtained from specialist drug agencies or through the local health authority. If appropriate, provisions should include plentiful free drinking water and a cool space (chill-out area).

795 Consider providing an emergency shelter or a 'crash marquee' which can serve as emergency accommodation for people who are left without transport home or without their own shelter at camping events. Such a marquee can also provide accommodation in the case of a major incident or contingency. In such instances, make sure the ground is covered to protect people from damp. If a marquee is provided, it will be necessary for the welfare workers to liaise with medical services

and stewards about access to these services. Sanitary facilities, emergency clothing or blankets, and refreshments may also need to be provided.

796 First-aid recovery/recuperation and observation can be provided at the welfare point if this is pre-arranged and supervised by medical workers, although it is recommended that cases that are more serious remain in the medical area.

797 Events provide unique opportunities for health education, particularly for young people. Wherever possible, encourage relevant health agencies to work in liaison with on-site welfare and information services on providing information on issues such as safer sunbathing, advice on drug use, safer sex, HIV/Aids and the provision of condoms.

Chapter 22

Children

798 Consider provision for children even if the event itself is not necessarily aimed at children, as they may accompany adults. Make sure that your publicity material indicates whether or not the event is suitable for children and if they are required to be accompanied by an adult, or if children under a certain age are not allowed entry.

799 The Children Act 1989 applies to 'activities' that run for more than two hours per day for six or more days of the year. Such activities will require registration. Children's activity organisers should consult the Registration and Inspection Unit of the local authority social services department for advice and clarification. Even where the Children Act 1989 does not apply, the Children Act provides sound guidelines regarding adult:child ratios and space standards for premises.

800 Consider the following matters in relation to children at the event:

- dedicated play areas;
- rides and activities which are not located in dedicated play areas;
- children with special needs;
- activities involving early teenagers;
- temporary arrangements for the safe care of lost children (which should be present at all events).

Planning and liaison

801 Consider the presence of children in your event risk assessment and major incident and contingency plans. The presence of pushchairs may need to be considered in evacuation plans. Ensure that there are suitable methods of communication with organisers of children's facilities and that other relevant on-site services, such as stewards, first aid and welfare, are aware of the provisions for children, their location, operating times, etc.

802 Ensure that the providers of services and facilities are aware of the possible needs of children and young people, eg bar workers need to be aware of the situation regarding the sale of alcohol and have methods of identifying those who are under age. Toilet facilities could include a 'mother and baby' room for baby feeding and nappy changing.

Care of children at dedicated play areas

803 Ensure that the areas are arranged and managed by people with relevant expertise and experience. Some key points are that:

- the designated leader of the children's play area has verifiable, relevant qualifications and experience;
- all helpers are over 18 years of age;
- people with known histories of child-related offences must not be involved;
- people involved are not under the influence of alcohol or drugs and must be medically fit;
- it is advisable for someone in the team to have a current first-aid certificate;
- people working with children know the nearest first-aid point and all emergency procedures;
- it is essential that all people working with children are fully briefed on all aspects of event safety policy and child protection issues;
- all accidents are recorded.

Child protection at dedicated play areas

804 Consider these matters:

- children under 8 years are not to be left unattended;
- a child must not be left under the supervision of only one worker;
- it is inadvisable for children under 8 years to leave the event unless accompanied by a parent or responsible adult;
- children are not to be taken away from designated children's areas, venues or sites unless parental permission has been obtained beforehand;
- children's workers are aware of the potential problems relating to the inappropriate handling or touching of children;
- children's workers are conversant with procedures for discipline and dealing with unco-operative children or their parents;
- corporal punishment (smacking, slapping and shaking) is illegal;
- practices that threaten, frighten or humiliate children are not to be used;

- dangerous behaviour by children is always discouraged;
- supervisors are aware of the problems that could arise from intruders and systems of communication with stewards are well established;
- workers should not undertake the care of sick children.

Facilities at dedicated children's areas

805 Make sure you have considered these matters.

- Is there adequate lighting, heating and ventilation?
- Is there supervised and controlled entry and exit at enclosed sites and venues?
- Are areas well defined or appropriately fenced?
- Is there provision for shade and shelter at open-air sites?
- The proximity to final exits in case of evacuation.
- There is no access to open water, eg ponds, for young children.
- Is there access to toilets and running water within a reasonable distance of the area or within the venue?
- Is there suitable food and drink for children?
- No smoking within children's areas.
- Toilets and washbasins are regularly maintained and cleansed.
- Is there suitable furniture for the age group of the children?
- Is there safety glass that conforms to the relevant British Standards provided at low levels?
- Are handrails at appropriate heights on stairs or steps?
- Floors are not slippery.
- Is there adequate provision for the regular and safe disposal of rubbish, including used nappies?

Numbers of children at dedicated play areas

806 The number of children in any area is based on the size of the space, the activities offered and the number of workers available. Where activities are provided for under 8 year olds, the Children Act 1989 may apply. Contact the local authority social services department for further advice.

Children's activities in dedicated play areas

807 Activities should be appropriate to the age of the children involved. Ensure that a risk assessment for each activity is carried out and everyone involved in the activity is fully briefed. Clearly display information about potential hazards. Pictures may be more effective for children.

808 Play equipment accessible to children must be safe and children need to be closely supervised when using the equipment at all times. Ensure there is constant supervision for activities that could be potentially dangerous, eg woodwork or candle making. Materials should be clean, non-toxic and non-allergenic. Particular care should be taken to check junk materials before use, for their cleanliness, suitability and any hidden sharp items, eg staples.

809 Face painters should only paint cheeks or hands of children under four years and always have parental permission. Ask face painters for their public liability insurance and proof that they are using a reputable brand of non-allergenic, water-based face paints. Provide information to parents about when and how to remove face paints.

Children of differing ages

810 Children's activity organisers should be aware that different age groups of children demand different levels of supervision and different types of activity.

- For children of 0-2 years, there must be nappy-changing facilities, qualified workers, suitable safe toys and all children must be registered on arrival. Provide hand-washing facilities and tables for supervising adults, and ensure equipment is regularly cleansed.
- For children of 3-5 years, qualified workers must lead activities. There should be access to a quiet rest area, activities should be varied and all children must be registered on arrival. The entry point must be supervised at all times to ensure that children do not leave the event unnoticed.
- Children of 5-7 years should be registered on arrival and a variety of interactive equipment and activities provided for them.
- For children of 8-11 years, registration may be advisable and more challenging activities that last for longer periods should be provided.
- For the 11-18 year old group, consider the situation regarding the sale of alcohol and methods of identifying those who are under age. Plenty of soft drinks should be easily available as an alternative.

811 If mixed age group activities are to be run, care should be taken with the layout of the children's area to avoid smaller children being knocked over or accidentally hurt by excited older children.

Children with special needs

812 Consider the requirements of children with special needs and develop a policy with regard to these children. Where events can provide activities for these children, provide suitable equipment and ensure that it is of a type that enables children with special needs to be fully engaged in the activity.

Rides, amusements and outdoor play equipment

813 Site rides and amusements in an environmentally child-friendly space. Rides should be appropriate to the age and size of users. Ride operators must have relevant statutory certificates for their equipment (see chapter on *Amusements, attractions and promotional displays*). Bouncy castle operators must provide proper supervision at all times. Fixed outdoor play equipment should conform to BS EN 1176 Parts 1-4, 6 1998, Part 7 1979 (replacing BS 5696).

Events involving early teenager audiences

814 Problems may arise with audiences of predominantly young teenagers who attend without their parents. They may be more prone to encountering difficulties, such as becoming separated from companions, missing the transport home and losing items, including money. Consider providing a 'help point' for these children.

815 Parents often take their children to these events, and sometimes have difficulties finding them again at the end of the event. Consider providing a specific meeting place for parents to wait along with a staffed message facility. An alternative would be to offer a space where parents can spend the time at the venue, waiting on site to collect their children at the end of the event.

Lost children

816 Prepare a 'lost child' policy that identifies arrangements for the safe care of lost children until they are reunited with parent/guardian.

817 There should be a clearly advertised collection point for lost children, supervised at all times with fully briefed workers. Ensure that lost children are not left in the sole care of a single worker (see paragraph 804). If there is a children's area on site, this will be the best place for the care of lost children. In some circumstances it may be necessary to make an announcement over the PA system. Take care to ensure that announcements do not refer to children specifically or give personal details, descriptions or names.

818 If a child is reluctant to go with a parent or collecting adult seek a second opinion from the police. The parent's or guardian's signature and proof of identity should also be obtained. Once a child has been reunited with their parent or guardian, inform stewards and police immediately if they have been involved.

Chapter 23

Performers

819 The requirements and responsibilities of performers have to be considered in event planning. Contract negotiations provide an opportunity to raise concerns and resolve safety issues in advance. Performers have responsibilities in relation to the safety of the audience and site workers. Performers could be held directly responsible for injury that results from their behaviour such as throwing things from the stage or not keeping to performance timings.

820 Supply the performer's management with a full briefing document before the event, including:

- how to reach the site and a map of the site showing specific artists' entrance, stage, stage plan and accommodation plan;
- an itinerary of what is happening, site access times, sound check times, performance times, etc;
- specific security arrangements.

Performers' areas and accommodation

821 Ensure that changing and 'warm-up' facilities are weatherproof, well lit and secure. Provide toilet facilities for male and female artists and consider separate toilet provision close to the stage.

Arrival and departure

822 Plan the arrival and departure times for performers. Their entry and exit points, if practicable, should be different to those used by the audience. Where risk of significant audience attention is

perceived, try to keep their vehicles out of view. Designate appropriate numbers of stewards to the area, if it is felt that performers will attract significant attention. Also consider the route to be taken to and from the venue. Some performers may arrive by helicopter so your risk assessment will need to cover the selection, marking and location of the landing zone.

Buses and other vehicles

823 Parking facilities for performers should, where possible, be separate from audience car parking and close to the stage. Where this is not possible, workers should be on hand, with appropriate transport if necessary, to help move people and equipment.

824 Ensure that the number of vehicles is kept to a minimum. Allocate a specific parking area for the vehicles, with the drivers available at all times in case they need to be moved. Many vehicles carry on-board generators and it is undesirable to keep these powered by leaving engines running. The vehicle operator should carry cabling to link up to a mains supply where possible. Where practical, consider providing a suitable mains supply (see *Electrical installations and lighting* chapter).

Workers and guests

825 Ensure that the number of workers and guests permitted into restricted areas is controlled so that the areas do not become overcrowded especially on stage and performance areas. Try to keep workers associated with performers to a minimum and ensure that they have suitable security clearance, which should be graded with access to key areas, such as dressing rooms.

Security of performers

826 Ensure that performers are met and booked in on arrival at the venue, suitable security passes are issued and where any threat, such as mobbing, seems likely, suitably trained stewards are employed. During the performance every effort should be made to secure the performance space. Artists and management should be made aware that they play a part in this process.

827 Advise performers of the evacuation procedures and the whereabouts of medical facilities. If this is not practicable, advise a senior representative who can shadow performers while on site, keeping in mind security needs and escape routes.

Performers' help in emergency planning

828 As well as being aware of the site safety arrangements, performers or their representatives can participate in the emergency procedures planned by helping to calm a situation and asking the audience to stand back (see paragraphs 269-278 on *Emergency public announcements*).

Chapter 24

TV and media

829 Music events attract a wide cross-section of media workers. Depending on the size, location and type of event, this can range from local media coverage through to global media attention. Large events can attract as many as 30 TV crews, 150 photographers, 200 journalists and up to 50 radio stations making a small community of up to about 500 people.

830 The management of TV and media can be split into two areas:

- pre-event;
- during event.

Pre-event

831 As an aid to crowd management and public information consider issuing a press release containing as much information as possible about the event: name, dates, times, location, line-up, ticket information, public transport information and contact name and telephone number.

832 Make sure that as well as national media outlets, all local media have been contacted with details of the event. If the event sells out or is cancelled or if a major incident occurs, good communications with local media will ensure that information is carried to the public quickly and efficiently.

833 Decide the amount of media that is manageable for the event. Setting an acceptable level of media attendance depends on how many people are able to look after them, how long the event lasts and available space.

834 All media can usefully provide advance advice to the public, such as conditions on site, travel arrangements, site facilities and restrictions. Ensure that each media representative who will be attending your site, receives information and advice on-site safety arrangements.

During the event

835 At medium- to large-scale events, consider setting up a small press tent or press office within the VIP or guest hospitality area (if provided). Ideally this should be situated away from production or artist changing areas.

836 The press tent or office is where information about the event can be posted, interviews organised and a meeting place set up for photographers, film and radio crews before media activity. If possible, provide the press tent or office with a payphone and power points so that the media can recharge batteries, phones, etc.

Photographers

837 Make sure that photographers are escorted into and out of the pit area and display appropriate passes. Where possible, photographers should enter and exit the pit area from the same side to allow medical services total access from the opposite side. If for any reason the pit becomes crowded or the safety of the audience is compromised the photographers should be made to withdraw. If there are a large number of photographers on site, it is recommended that they should be escorted to the pit area in smaller manageable groups to prevent overcrowding of the area.

Radio broadcasters

838 Local radio stations often attend the site with a mobile or outside broadcast unit (OB unit) to feed live inserts or soundbites back to the studios. The OB unit usually takes the form of an estate car or four-wheel drive vehicle with a large telescopic mast. Once the interviews are completed ensure the OB unit is moved off site or to allocated parking areas.

839 Sometimes a radio station is set up on site specifically to broadcast programmes to the audience. Not only does this provide entertainment and interviews with performers and the audience, but also it can be extremely useful in transmitting important safety information and messages for people. Plan how you will access the radio station with safety information you may wish to be transmitted.

Press journalists

840 Press journalists normally require the least amount of attention as they attend to review the event as a whole rather than acquire individual interviews. Once again, however, ensure that any interviews with artists are pre-arranged before the event begins to limit the amount of on-site organisation between press office and artists.

TV broadcasters

841 TV requires the most amount of attention and the type of TV crews and workers can be broken down into three main areas:

Event filming units

842 A large event will generally be recorded by a dedicated production company or broadcaster for live or future broadcast. Plan these arrangements in advance as they require special facilities such as filming platforms, OB vehicles parked backstage, audio and visual mixing trucks, front-of-house filming platforms, etc, these arrangements need to be considered in venue and site design.

TV news crews

843 TV news crews will consist of local news crews, cable and satellite crews. These crews are normally small (two to four people). They will require only a short amount of time on site and therefore can be serviced relatively quickly. They should be supervised wherever possible and escorted quickly and efficiently to key locations (production offices, services offices, front-of-house, etc).

Production companies

844 This is normally the largest part of the TV mix and is made up of music programmes interested in covering the event, to lifestyle programmes.

845 TV crews tend to need access to vehicles for equipment and storage, so consider space allocation close to the hospitality/VIP area as possible. Other non-essential vehicles can be allocated spaces in designated car parks.

Foreign media

846 Foreign media workers need to be clearly briefed in advance and given assistance to understand the safety requirements especially regarding the provision, compatibility and use of electrical equipment.

Student media

847 Student media can be helpful to the event and useful for communicating with the audience at events attended by predominantly young people.

On-site structural considerations

848 In addition to requirements already mentioned for facilities, vehicles and accommodation, the presence of media workers, and TV broadcasters in particular, will have to be considered in your venue and site design. Media provision such as gantries may restrict viewing areas for audiences and therefore cannot be counted in occupant capacity calculations.

849 Media may need to use constructions, as outlined in the chapter on *Structures*, such as scaffold towers and the use of barriers around media installations. Similarly, requirements for electricity supplies will need to conform to recommendations, eg burying cables.

On-site public relations staffing requirements

850 The number of workers required to manage media will vary according to the size of the event, the number of days that the event is held over, the type of event, the capacity and the amount of media expected. At large three-day music events with a 50 000 or more capacity, at least ten people will be required to deal with the media. At smaller one-day events, four to six people should suffice.

851 Issue all workers with site radios that have their own dedicated channel to avoid taking up unnecessary time dealing with media questions, guest lists, artists' whereabouts, etc. Ensure that all media workers are fully briefed and aware of any emergency procedures.

852 Media liaison workers need to have a base; normally this will be at the point where media representatives check-in or at a press tent. These areas are best sited close together to avoid large distances between the production areas, pit, press-tent and media check-in locations.

853 Over the course of an event, media and press officers will get to know the individuals involved and this is extremely useful in the case of emergencies or important announcements. Ensure that the chief press officer is introduced to key stewarding workers, local authority officers, police spokespersons, welfare organisers, event film units, etc. This will enable a direct line of communication between the media and services that is controlled and efficient.

Chapter 25

Stadium music events

854 Staging events in stadia differs from other venues. This chapter begins by describing the general organisational arrangements, legal issues and management responsibilities and is followed by practical matters to consider when staging a music event in a stadium.

855 Sports stadia are increasingly used to hold events other than those for which they were originally designed. Music events will normally require a public entertainment licence. However, in the case of sports stadia additional legal requirements will apply under the Safety of Sports Grounds Act 1975. If the stadium is designated under the 1975 Act, eg league and premier football grounds and national stadia, a general safety certificate will have been issued, but will be unlikely to cover activities such as music events. A special safety certificate will therefore need to be issued for a music event.

856 In some areas of the country the entertainment licensing and safety certification responsibilities may not be held and administered by a unitary authority but split between local authorities at district and county level. Where two local authorities undertake these functions it is essential that there is close liaison between the authorities to ensure consistency in safety standards and requirements.

Guide to safety at sports grounds

857 Sport stadia regulated and certificated under the 1975 Act are assessed for occupant capacity, safety, comfort and welfare of spectators by standards contained in the *Guide to safety at sports grounds*. When a sports ground is used for a music event, the information in the *Guide to safety at*

sports grounds and this publication will apply. The provisions of this publication will be necessary to supplement those of the *Guide to safety at sports grounds*, either in respect of the matters not covered, or that require an alternative approach because of the different circumstances.

858 It is very important when deviating from the recommendations of the *Guide to safety at sports grounds*, to ensure that safety decisions are recorded and justified with written evidence. An event specific risk assessment will need to be carried out. These matters are the responsibility of the stadium management team.

Management issues

859 The stadium management will be the certificate holder under the 1975 Act and will remain responsible for the provision and maintenance of safe accommodation and safety standards for an audience coming onto the premises. It is also quite likely that the local authority will grant the public entertainments licence to the stadium management as occupier of the premises. However, it is equally likely that the promotion and staging of a music event will be undertaken by an outside organisation.

860 To avoid conflict between the parties and blurring of lines of responsibility for specific safety duties, close liaison is essential together with a record in writing of the agreement between the parties regarding the responsibility for safety or safety functions.

861 Stadium management need to ensure that the proposed event, including temporary structures, will fit into the stadium and that all proposed audience accommodation, both permanent and temporary, is safe and sufficient for the use intended. Also consider existing structures and whether they can cope with the dynamic loading likely to result from the movements of a concert audience as opposed to those of sporting fans. Further information can be found in the Institution of Structural Engineers document *Temporary demountable structures: Guidance on design, procurement and use.*

Planning for safety

862 Ultimate responsibility for the safety of a stadium audience rests with the stadium management and not with the event organiser, unless they are one and the same. Stadium management therefore need to have a safety management structure in place and be able to demonstrate their compliance with health and safety legislation.

863 In particular stadium management should have:

- a written safety policy for employees and audience members;
- contingency plans;
- agreed major incident procedures;
- agreed statement of intent documents;
- written risk assessments that justify actions, activities, plans and methods of operation.

864 Stadium managers will or should be actively involved in safety advisory groups, which include officers of the local authority and emergency services. Such groups are useful for assisting and advising the stadium management on how to discharge their responsibilities and should be contacted at an early stage during planning for music events. This assistance however, does not relieve stadium management of their responsibility towards people visiting their stadia.

865 Start safety planning at an early stage, as much of it will be critical to important issues including occupant capacities. It would be wise to review existing plans and risk assessments as these may still be appropriate under the temporary arrangements or may only need adjustments or modifications to take account of the different circumstances.

866 Some safety plans may need to be re-written to deal with specific situations. For example, planning the evacuation of audience members from terraces, which may involve an element of forward movement onto the pitch as a place of safety. Obviously if the stadium is being used for a music event, with the pitch area designated for the audience, the likelihood that members of the audience will try to evacuate the stadium via the terraces must be catered for.

867 Those involved in the process of reviewing, revising and redrafting major incident and contingency plans, risk assessments, etc, will benefit by walking the site. This simple exercise will prove extremely beneficial and highlight issues not otherwise thought of that might have the potential to cause harm and pose a risk.

Audience size

868 The means for calculating the audience size or occupant capacity for an event is dealt with separately in this publication. However, calculating the occupant capacity for an event within a stadium will require a range of additional factors to be considered. The purpose of setting the occupant capacity is to ensure safe entry, safe accommodation during the event, safe exiting at the end of the event and that the means of escape are adequate for the number of people attending the event, and in the case of music events with standing provision, it also assists in controlling audience densities. There is also a benefit in knowing the maximum capacity when determining the level of provision needed for first aid, stewards, toilets, catering facilities, etc.

869 When calculating the occupant capacity for a sports stadium certificated under the 1975 Act, there is the immediate advantage of already knowing the set capacities for the normal spectator areas, which will also be provided with adequate means of escape. However, if the pitch area is occupied by the audience (standing or seated) and/or by temporary structures, such as a stage or stands, additional exits are likely to be required. Whatever the assessed capacity and density is, it must be emphasised that the final capacity should not exceed the available calculated means of escape. It would, therefore, be sensible to initially calculate the occupant capacity based on the exit potential of the stadium.

Factors affecting the occupant capacity

870 The additional factors that will influence and reduce the final occupant capacity due to loss of space and restricted viewing include:

- those areas of the stadium closed to the audience, eg back stage;
- sight-line obscured areas, eg stage left and right next to back stage. If the stage is deep and the performers work from the back of the stage, the poor sight-line areas particularly in the terrace seating will increase;
- stage size and location on terrace and pitch - if the stage is large and positioned too far forward, valuable audience space is lost;
- those areas taken up by the sound and lighting mixer unit, delay towers, camera platforms, catering and merchandising stalls;
- reduced terrace seating removed to allow for wider gangways onto and off the pitch area;
- satellite stages and large thrusts for 'acts' to perform, together with protected access routes;
- multiple barrier systems and pens - when these systems are used to the front of stage, a considerable amount of pitch area is taken up by the barrier construction and sterile gangways for the stewards and first-aid teams;
- first-aid points located on the pitch;
- areas both on the pitch and in the terraces that are in the shadow of structures - such as the mixer tower. These positions often have very poor sight lines and as such are not occupied;
- migration of the audience down from the terraces onto the pitch - this displacement can lead to higher audience numbers and density on the pitch than considered safe.

871 The location of the stage structure in the correct position as agreed on the site plans cannot be over emphasised. It is essential to ensure that a competent person oversees the stage installation into its correct position. If this does not happen and an error occurs with everything being 1-2 m further forward onto the pitch, many safety-related problems will ensue, eg many terrace seats will suddenly have poor or no sight line of the stage, the pitch area available is reduced (increasing the audience density), and the pitch structures will correspondingly have to move, affecting more audience sight lines - remember that this will all happen at a point in the event production when the bulk of the tickets will have been sold.

Access management

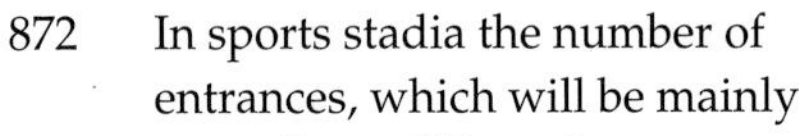

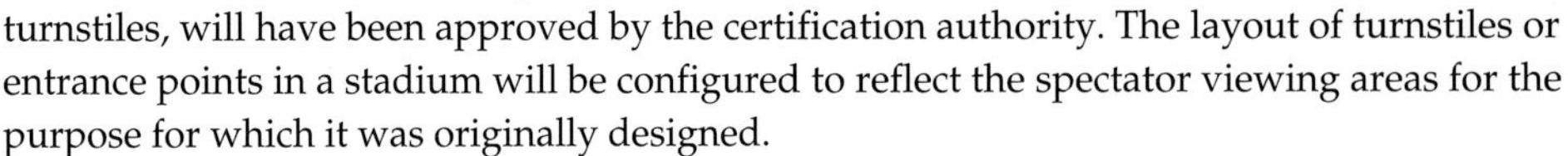

872 In sports stadia the number of entrances, which will be mainly turnstiles, will have been approved by the certification authority. The layout of turnstiles or entrance points in a stadium will be configured to reflect the spectator viewing areas for the purpose for which it was originally designed.

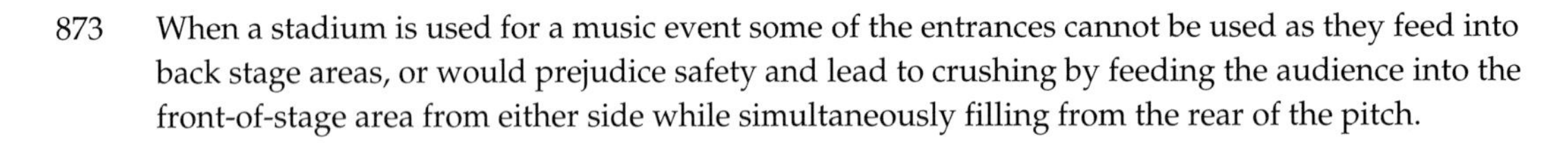

873 When a stadium is used for a music event some of the entrances cannot be used as they feed into back stage areas, or would prejudice safety and lead to crushing by feeding the audience into the front-of-stage area from either side while simultaneously filling from the rear of the pitch.

874 If all entrance points are not available, it will be necessary to determine how the stadium can be loaded safely for the assessed occupancy capacity. The standard for sports stadia is based on a maximum notional flow rate of 660 people through a turnstile in 1 hour or the measured flow rate, whichever is the lower, with an overall ground-loading time of 1 hour (this standard reflects the arrival pattern of football spectators). A longer loading time with a reduced turnstile flow rate may be acceptable as long as the safety of the people queuing to enter will not be compromised.

875 Consider the following factors:

- the profile and nature of the audience and likely or known behaviour;
- assumed slow arrival over much longer period;
- entrances opened at least two hours before the first act;
- the actual number and location of turnstiles proposed to be used;
- the level of security checks of the audience coming through the turnstiles;
- whether the event is a general access show, ie free movement of people;
- plans to feed pitch from opposite end to stage;
- seated pitch - this reduces need for one end access and allows for greater number of turnstiles to be opened;
- reserved seating - encourages people to turn up later and not rush access;
- ticket sales on day (should be remote from turnstiles, as should ticket pick-up points.

876 Crowd conditions can deteriorate outside of stadia if large numbers of people seek to arrive and gain admission at the same time - this is a particular problem with evening events when access points simply overload and cannot process the audience members into the stadia quickly enough. Health and safety responsibilities extend to people outside the stadium, even if people are on the public highway, so consider the following matters:

- early consultation and planning with the police and provision of stewards for marshalling the queuing audience;
- suitable barrier systems are available if required;
- adequate numbers of properly trained stewards are deployed;
- a PA system needs to be available and can deliver announcements to the waiting audience (keeping people informed of events and timings is reassuring and settling);
- use of CCTV to monitor queues;
- catering facilities are available if they can be located so as not to cause an obstruction, restrict queuing or free movement (if on the public highway it may be necessary to discuss this with the local authority);
- first-aid facilities to the outside of the stadium are essential;
- toilet facilities are available - it may be necessary to provide mobile units (this should be discussed with the local authority if on the public highway);
- contingency plans are in place to deal with audience problems and in particular the relief of pressure points in front of or near to turnstiles.

Exit and evacuation arrangements

877 Spectator exit and evacuation arrangements will already be in place and approved for the stadium. However, for music events audience accommodation will differ from the standard arrangements. Consider issues relating to the accommodation of audience members on the pitch, whether seated or standing and the requirements for additional exits that do not deliver people into occupied terraces. Evaluate the need to evacuate people from the temporary structures. If multiple barrier systems or pens are used to the front-of-stage, careful planning and recording is needed. Plan how the audience will exit without delays, and ensure that any barrier systems will not affect exit routes from the pitch.

878 Emergency evacuation plans will have to be revised or re-written in respect of the terraces as standard plans that may include the use of the pitch as a means of escape or refuge, will not be available.

879 When considering exit and evacuation strategies, particularly if a stadium is being used for the first time for music events, the audience will respond in a different and slower way to instructions, controls and emergency evacuation procedures. There will not be the same sense of urgency to leave the ground that regular sports fans will have.

The pitch

880 Music events in stadia that include the use of the pitch for audience accommodation must be safe to use with associated access to facilities. Arrangements should include:

- the provision of firm under foot conditions, that do not present tripping hazards in themselves and where they join other surfaces and levels, are not slippery and cannot be set alight;
- specific provision, such as tunnels or custom-built pitch cover units or tiles that incorporate cable runs that link the mixer unit with the stage facilities;
- barrier systems that protect temporary structures on the pitch;
- planned interface of temporary structures and barrier systems with the terracing to avoid unnecessary sight-line obstruction, exit obstruction and slowing of free movement to facilities;
- drinking water points strategically placed around the pitch perimeter that do not create pinch points or obstructions and water-logged areas - these should be effectively stewarded;
- access points onto and off the pitch through the terraces to allow free movement to internal facilities, shelter and rest from the activity. It is critical to keeping good order and movement that one-way systems are introduced using the terrace gangways, which should be controlled and monitored by stewards and ground CCTV. These points near to and on either side of the front of stage must only be used as exits off the pitch to prevent overcrowding and potential crushing at the front of stage;
- gangways off the pitch must be kept clear of people trying to gain an improved view.

Audience profile - service implications

881 Music events held in sports stadia attract different groups of people to traditional sports events. The audience profile will vary, attracting more women, children, people with special needs or elderly people and larger groupings of particular generations. The difference in audience profile taken together with considerably longer event and loading times, will require careful evaluation of the suitability and adequacy of services available or planned for the occupant capacity. Thought needs to be given to the way in which the audience members will use or want to use the premises, and the need to revisit, revise and adapt contingency plans in the light of the above.

First aid

882 Some of the normal ground facilities may be in areas that are not open to the audience for the duration of the event. Additional facilities will be needed on the pitch, especially either side of the stage to assist with those people taken over the front-of-stage barriers. When the pitch first-aid stations are planned it is important to locate them so that they are not exposed to excessive sound levels from nearby speaker stacks, and access for stretchers and trolleys is good - don't forget the provision of power for lighting and water. First-aid provision will also be required to the outside of the premises and greater numbers of medical staff may be needed to cope with the longer duration of event.

Sanitary facilities

883 The current stadium facilities will have ratios of males to females that will need adapting if there is an increase in the female audience numbers. When considering the location of these units, service arrangements, ie sewerage and water connections, must be available and their location must not obstruct or block circulation routes, especially those affecting exits and evacuation. When calculating toilet provisions consider the likely weather conditions and the drinking capacity of the audience.

Signage

884 It is likely that some people attending music events at a stadium will be unfamiliar with the layout of the premises, which together with temporary changes to the usage of the building will require additional signs backed up by stewards at key points, eg one-way access points onto and off the pitch. These extra signs must comply with the Health and Safety (Safety Signs and Signals) Regulations 1996 and not conflict with existing signage which may have to be either taken down or covered up.

Lighting

885 Evening shows and events are likely to have reduced 'house' lighting levels to the pitch and terracing areas to enhance the special effect and stage lighting. It is essential that emergency lighting and exit lighting is maintained and not obscured. The pitch area must have a suitable level of lighting available for emergencies and at the end of the event when the audience leaves. Sole reliance on the stadia floodlights for this purpose is not acceptable, as these systems do not respond instantly taking time to power up - alternative supplementary lighting should be available.

Communication

886 It is essential that the stadium PA system can cut into and interrupt any imported show sound systems so that the stadium management's event control can relay safety announcements. Ambient noise levels inside a stadium during events can be high, unlike those experienced at normal sporting events. Support services such as security, stewarding, medical and fire safety workers rely upon radio communications. It will therefore be necessary to ensure that the systems can function within the environment. This may require upgrading of microphone and earpiece kits for portable radios and changes to operating methods.

Electrical installations

887 It is important to stress that the normal systems incorporated within the stadium will not be sufficient to meet the power demands of the show production for stage effects, lighting and support service. In addition there will be further power demands for the temporary catering and toilet units installed within the venue, which also could overload in-house circuits if not provided with an independent supply.

888 The normal source for the additional demands are generators. These must be installed and operated by a competent person. Their location should not obstruct exit routes from the stadium and should not create pinch points in audience circulating areas. Generators should also be secured by barriers to prevent unauthorised access. Cable runs from generators should be carefully planned and monitored so that they do not obstruct the safe movement of people, and that cables are not exposed to damage from vehicles, fork-lifts trucks, etc.

Production facilities

889 Stadia usually have minimal administrative and support facilities that will be used by stadium management and the emergency services on event days. However, the requirements for extra office accommodation, storage, crew catering, plant, equipment and lorry pounds, and artists' quarters, will be considerable. These 'villages' will develop and usually grow from the edge of the building backstage. These facilities while critical to the show, must be planned, monitored and secured from the audience ensuring that their position does not interfere with pedestrian access and exit routes or emergency services vehicle paths.

Structures

890 It must be stressed that responsibility for the safety of temporary structures erected within the stadium rests with the stadium management and reference should be made to the chapter on *Structures* in this publication for detail. It is worth reminding all those responsible within a stadium environment for the erection of temporary structures that there are other serious safety implications which arise from their poor positioning in relation to proximity of other structures, exit, access routes and facilities. The potential for dangerous audience movements can occur due to poor positioning with funnels and pinch points being created, and narrow gaps between protective barriers resulting in high audience densities.

Special effects

891 Fireworks displays, often at the end of a show, should be arranged so as not to fill the stands with smoke, and shower debris into the faces of the skyward-looking audience inside the stadium and public outside. Such firework displays are often by necessity fired from the stadium stand roofs, which do not always afford safe access and by their nature are fragile in construction.

892 Many special effects used in shows are custom designed and built for specific event tours and may include the use of fuel such as bottled propane. The use of bottled gas is not normally permitted. However, approval may be given to one-off effects if the organisers and stadium management can justify its presence on the grounds of alternative safety factors.

Event workers and facilities

893 Music events held in stadia will draw in many workers who are unfamiliar with the particular ground and its safety management arrangements. It is, therefore, essential that stadium management ensure that all those coming into the stadium - such as technical staff, contractors, contract stewards and security, media personnel, promoter's staff, officials, participants and artists' representatives - are adequately briefed and familiarised in advance together with the chain of command and clear split of responsibilities.

894 If stadium management are making arrangements for stewarding provision and are likely to be using the same pool of resources as used for sports events, the stewarding training may need to be re-assessed. Account also needs to be taken of the stewards' welfare due to the longer duration of the event, and the need for catering and rest facilities. At music events there are some specialised tasks, such as stewarding the front-of-stage pit that require additional skills, not normally held by the resident stewards and this factor needs to be taken into account.

895 For long-duration events, also consider the 'house keeping' issues regarding the significant build-up of rubbish and pressure on all facilities including toilets. It is essential that adequate workers are available to deal with these matters and keep the facilities safe and clean. The disposal of the large amounts of refuse generated at music events require pre-planning so that audience safety is not jeopardised.

896 Due to the nature of music events and the early arrival of the audience, the event control centre/room will need to be operational earlier than is usual and be prepared to operate through to the end of the event.

Chapter 26

Arena events

897 This chapter aims to highlight some of the factors to consider when organising a music event in an arena-type environment. An arena can be defined as an indoor area used for the purpose of public assembly. This definition can therefore cover a variety of different premises ranging from purpose-built arenas specifically designed for the holding of music events through to premises not originally designed and built for the purpose of holding music events.

898 Many arenas are multi-functional so as well as hosting music events they will also be likely to host fashion shows, sporting events, exhibitions and conferences. In larger complexes, with several different sized arenas, it is possible that more that one event could be in progress at any one time. This chapter is specifically designed to give advice about music events although there may be elements of good practice contained in this chapter that can be applied to other type of events in arenas.

Planning and management

899 Arenas specifically built for the purposes of holding music events will almost certainly have an annual entertainment licence, which will be held by the arena owner or manager (referred to as the arena operator for the rest of this chapter). If you wish to stage your own event in an arena already holding an entertainment licence you will need to liaise directly with the arena operator. The most important planning aspect will be determining the responsibilities for health and safety between the respective parties and documenting the agreements.

900 Some arenas will not have obtained an annual entertainment licence or the arena operator will require you as the event organiser to obtain an 'occasional' entertainment licence specifically for the event in your own name. In these circumstances, it will still be necessary to determine health and safety responsibilities especially if other events are taking place in different parts of the premises.

901 Arena operators will already have in place a written safety policy, risk assessment and major incident and contingency planning documents required for their own workers and events they organise themselves. If you are hiring an arena (or part of it) and/or obtaining an entertainment licence in your own name you will also need to liaise with the arena operator so that information about the existing safety management systems can be exchanged.

902 Health and safety responsibilities need to be determined for the preparation of the risk assessment for the music event. Agreements also need to be documented on the services supplied by the arena operator including workforce and equipment. There may be a need to nominate a safety co-ordinator, from one of the parties.

903 Arena operators are likely to have prepared their own in-house safety procedures which need to be communicated to any external contractors brought onto site. In multi-occupied premises it is important that agreement is reached and health and safety responsibilities assigned between the parties, in relation to major incident planning. Ensure that the planning for the event is co-ordinated with the planning of the premises as a whole.

904 A system to ensure that health and safety information is communicated to other users of the building, especially if there is more than one event occurring at the same time, also needs to be documented and agreed.

905 The breakdown of the event may have to take place very quickly if the arena has been booked for other events. Working to tight deadlines needs careful planning to avoid creating tired and stressed workers and contractors more prone to make mistakes.

906 All these matters should be discussed at the initial planning meetings with the local authority at the event safety planning meetings. It is recommended that arena operators and event organisers should follow the guidance set out in the *Planning and management* chapter.

Crowd management

907 Some arenas require the existing stewarding and security staff to be employed, others do not. If external stewarding or security contractors are to work alongside the existing stewards, clear lines of control and co-operation need to be established. All stewarding and security workers should operate through a central control. The role and duties of security and stewards should also be clearly defined.

908 Arenas not originally designed for music events, especially located in the heart of cities and towns, are unlikely to have adequate queuing areas. Members of the audience may arrive a few hours before the official opening of the event. In these circumstances it will be necessary to plan the

provision of barriers to prevent audience members queuing onto the public highway. Stewarding and communication systems may need to be considered in these circumstances to keep the audience members informed and relay any special safety information.

909 Large amounts of litter can be generated in queues including glass bottles and cans. Extra waste receptacles may be needed. You may need to provide toilet facilities outside the arena.

910 Management of the audience arriving and leaving the arena should be discussed with the police, highway authority and local authority. Extra stewards may be needed to direct the audience leaving the venue as to where to find waiting coaches. The majority of the audience members will leave the event at the end. It is important that there has been proper consultation with the public transport providers to ensure that sufficient public transport is available (see chapter on *Transport management*).

911 Agree with the police mechanisms to control unruly behaviour outside of the arena. It is usual that there will be a staggered arrival of audience members to the arena. This may, however, be dependent upon the allocation of numbered seated events as opposed to standing events. In the latter case audience members may wish to arrive early to secure a standing position close to the performance/stage area.

912 Make sure your major incident and contingency plans take account of the problems caused by the need to evacuate the arena. The sudden influx of the audience into surrounding streets could cause traffic congestion problems and prevent access for emergency vehicles. It is important that the audience are informed as to whether the event will continue or if it will be cancelled. This will determine whether the audience members stay or leave the area to make their way home. Both scenarios should be planned and stewards suitably trained in the necessary procedures.

Transport management

913 Information concerning the availability of car parking, public transport and other forms of transport such as special bus or coach services should be advertised with the music event or printed on the tickets. Whether the audience arrives by car, coach, train, bus or underground train will be very much dependent on the availability of adequate parking areas and the public transport facilities near to the arena (see chapter on *Transport management*).

914 Examine transport management in relation to the major incident planning for the event. If the arena needs to be evacuated and the event cancelled after starting, plan to take account of the sudden and unexpected influx of the audience members on to the public transport network and surrounding roads. It is important that the transport providers, police and other agents receive information quickly to enable them to put their emergency procures into action rather than react to the situation as it occurs.

Venue design

915 When planning arena events, remember that there is a certain degree of inflexibility compared to greenfield sites, eg the size of the premises, existing WC facilities and fixed entry and exit points. The occupant capacity of the arena will therefore be primarily dependent upon the means of escape in case of fire, the limiting factor being the width and suitability of the exit doors for the different standing/seating configurations. To ensure that the arena is suitable for the music event it may also be necessary to bring in temporary equipment, such as extra toilets and electrical generators.

916 Arena operators need to agree with the fire authority and the local authority the different standing/seating configurations that can be used within the arena. Event organisers can then be supplied with a copy of the various acceptable arrangements. Prior approval of specific arena layouts can be useful.

917 The position and design of any necessary barriers should also be taken account in the overall design of the arena. The positioning of all structures, no matter how small, should be shown on the plans, eg food concessions and display stands as these can have an effect on the safe evacuation of people in the event of fire or other emergency.

Structures

918 Arena operators may have their own facilities and staff to erect the necessary structures for a music event, eg stages and seating. You may, however, need to install your own structures. Agreement should be clearly documented as to what structures and other equipment will be brought into the arena and who will be responsible for its correct positioning, safe erection and use. Incorrect positioning of stages can have a serious effect on viewing areas (see chapter on *Stadium events*).

919 Health and safety management systems between an external workforce brought onto site and the existing internal workforce should be defined and documented. You will need to ensure the competence of external contractors brought onto site. The existing health and safety procedures should be brought to the attention of the contractors.

Chapter 27

Large events

920 For the purposes of this chapter a 'large event' normally has one or more of the following components:

- multi-stage;
- multi-performance;
- multi-activity;
- multi-day;
- physical size of venue (outdoors).

921 The significant factor is, however, the size of the audience - commonly 15 000-35 000 but sometimes in excess of 100 000. It would be easy to regard a large event as being the same as any event but with more facilities, services and workers, etc. While reference should be made to the specialist chapters of this publication, there are a number of areas where the size of the event demands particular attention.

Planning and management

922 The need for extensive consultation and planning cannot be overemphasised. The formation of an event safety management team, comprising representatives of the emergency services and local

authority, is a useful method of addressing the practicalities of event organisation. Team meetings can be scheduled before, during and after the event and can run in parallel to any formal public entertainment licence procedure. Given a sufficient lead-in period, it should be possible for the safety management team to develop into a working unit that can resolve any difficulties.

Crowd management

923 While the proposed attendance figure is the key to the provision of services and facilities, account should be taken of the number of guests and staff. Dependent on the event, up to 10% of the capacity could be guests or staff at the event with the consequent additional load on site infrastructure.

924 Also consider easing local traffic congestion by opening the site early and restricting exits. Incremental occupation of the site should be accompanied by a similar incremental provision of services.

925 In some instances for nominally non-camping events, it may be useful to make contingency camping provision and low key entertainment on a normally silent night. There is, however, a danger of changing the nature of the event for subsequent years. Ticket pricing structures may control arrival, particularly in terms of late Friday arrivals for a Saturday event.

926 Within the site there needs to be active crowd management. The technical issues of stage layout, audience size and barriers are dealt with elsewhere. At a large event the layout should take account of audience movement across the site and minimise cross flow and points of congestion. Ideally a wheel layout, with entertainment at the hub and camping at the rim, could be combined with one or more of the following:

- area or cellular stewarding to maintain a controlled scale of audience movement;
- dynamic entertainment management where the programmes on separate stages are integrated into the audience management programme;
- ensuring that timing and running orders are adhered to, to avoid conflicts at the end of performances;
- gradual close down of main stages;
- continuing (perhaps for 24 hours) low level entertainment such as cinema or markets;
- no entertainment within the defined areas of campsites.

Major incident planning

927 The size and complex infrastructure associated with a large event reinforces the need for a comprehensive major incident plan. The event safety management team in consultation with the local authority emergency planning officer, who would know about local arrangements, should develop the plan. The following aspects should be considered.

- Is the evacuation of the entire site practical or would selective evacuation be preferable?
- Is the evacuation of the site desirable, given that under some circumstances food, water and sanitary facilities may still be operational on a scale not available elsewhere?
- What infrastructure is available elsewhere?
- What would be the impact of a mass exodus from one part of the site on other parts or on the locality?
- What implications are there for public address systems in various emergency situations?

Transport management

928 If public transport links are available they may be encouraged by the use of integrated ticketing. Depending on the event and availability, many people may choose integrated coach/event travel.

In rural locations, or where other transport is unavailable, a high proportion of the audience will, almost inevitably, travel by car and the logistics and impact on the locality should form an early item for consultation.

929 Traffic should be removed from the public road system onto site, as quickly and efficiently as possible - the use of professional stewarding may be the best option. Within the site, parking areas should be divided into easily identifiable zones (perhaps associated with nearby camping) and traffic should be routed to avoid designated pedestrian routes/areas.

Children

930 A function of the size of the event is that people may become more easily lost. The need for overnight provision and the implications of the Children Act 1989 need to be considered (see chapter on *Children*).

Information and welfare

931 The provision of a comprehensive information and welfare service that can assimilate and co-ordinate information in an active way as the event progresses allows other agencies, such as medical and police services, to undertake their specialist functions. In general terms, everything possible that an individual requires for the duration should be readily available on the site. It is particularly important that there are a sufficient number of food stalls to cater for the audience demands.

TV/media

932 The presence of national and international media may in itself influence the progress of an event. In particular, incorrect ticket availability broadcasts may cause problems. Ensure that channels of accurate information are available for co-ordinated release to the media.

Venue and site design

933 The site design for a multi-day event must recognise the need for 24-hour access to facilities for both the audience and for servicing the facilities.

Fire safety

934 Discussion should take place, pre-event, on the areas of responsibility for fire safety teams. There needs to be a clear understanding of the circumstances under which the local fire brigade will attend and lines of communication must be established. There should be a policy and procedure in place for safely dealing with small arena fires.

Sanitary facilities

935 The availability of water is a limiting factor on the audience size at all greenfield events. In particular, the logistics of moving large quantities of liquid, whether water or effluent, need to be addressed. While flush toilets are a preferred option, they are vulnerable to failure of water supply and can be very difficult to bring back into use when the supply has been restored. The use of fewer toilet blocks with more units can, subject to proper access routes and efficient continuous servicing, mean that a greater number of toilets will remain in operation. For overnight or multi-day events, there will inevitably be a peak morning demand, particularly if showers are provided in camping areas.

Food and drinking water

936 Supplies of both food and drinking water must be adequate for the duration of the event; the facility for campers to buy basic commodities such as bread, milk, etc, needs to be available. To ensure sufficient supplies of water there will need to be a considerable amount of temporary pipework, which is susceptible to damage and vulnerable to contamination. Consideration should be given to splitting up the water supply on the site into several independent supply zones. In this way the consequences of a serious incident affecting the water supply will not affect the whole site. It may be necessary to protect the quality of the supply by increasing chlorination above normal mains levels. The use of percussion taps will help reduce waste.

Health and safety of event workers

937 Set up a proper management infrastructure with delegation of responsibility. The safety management team should include people with experience of previous or similar events. One of the issues that will be encountered with large events running over many days is one of fatigue among both management and contractors. All will be working long hours under stressful conditions and if this is not addressed, the quality of decisions, some of which may be critical, could be poor.

Chapter 28

Small events

938 This chapter contains advice aimed at the small-event organiser. The immediate difficulty is defining what is a small event and following on from that decision which parts of this publication apply and which parts do not. The important factor to consider is not whether an event can be defined as 'small' or 'large' but the level and extent of facilities and safety management systems required at your event to ensure the health, safety and welfare of the people attending.

939 The advice in this publication has been written for events with over 2000 people attending. The safety and welfare recommendations therefore reflect this figure. For the small-event organiser this publication can help you think about the safety matters to be considered. Your overall event risk assessment will help to determine what systems or precautions you need to put into place to manage the event safety. Remember, however, that managing any size of music event will require good safety planning procedures.

940 While this chapter covers small events taking place either wholly or partially in the open air or in marquees or other temporary structures, it may also contain useful safety advice for small events taking place indoors in fixed buildings.

Planning and management

941 It is suggested that small-event organisers use the chapter headings in this publication as a framework or checklist for event planning. All event organisers must be clearly aware of their responsibilities for the audience and other participants at their event, including performers, traders, etc.

942 Small-event organisers should not assume that because a proposed event is deemed to be small, the associated risks are less. Not only will the number of people attending be significant for the event management, but the activity itself and the audience type will also influence the safety requirements. It is just as important for a small-event organiser to carry out a risk assessment for the event, to identify which hazards are of greatest significance and therefore which parts of this publication are of most relevance. Simple hazards on greenfield sites, such as rabbit holes, old barbed wire in long grass, the presence of recent animal droppings, etc, are as much a danger to a small audience as to a large one.

943 A safety policy statement should be produced that describes how the event organiser intends to manage safety; who has specific responsibilities; and how these will be carried out. The risk assessment and safety policy need not be long or complicated, but should clearly demonstrate the approach taken to ensure the safety of all those involved in the event. Assistance in drawing up a risk assessment and safety policy can be found in the *Planning and management* chapter and in the HSE documents listed in the *Reference* and *Further reading* sections.

944 A safety management team should be formed to put the actions outlined in the safety policy into practice. Two to three people would be sufficient for a small event. A list of site safety rules should be drawn up and distributed to all workers or helpers who need to be aware of safety procedures. Ensure that any contractors or subcontractors hired to build the stages, erect marquees or stalls, etc, are competent in managing their own health and safety on site. Ask for copies of the contractors' safety policies, risk assessments for their work and safety method statements.

Staffing

945 Small events may operate with small budgets and rely on enthusiastic helpers rather than paid employees or contracted service companies. The crucial aspect is good co-ordination by the event management team and close supervision, support and monitoring of helpers. The organising group can sometimes provide many services at small events such as catering and stewarding, rather than buying them in from commercial companies. All helpers will need to be aware of legislation, regulations and guidelines affecting the provision of services.

946 Management of workers and helpers requires clear job functions and responsibilities to be identified. It is particularly important for inexperienced workers and helpers to receive proper training and supervision.

947 Everyone working or providing services at the event should be clear about what they are required to do, how to do it and when it needs to be done. This can be achieved by preparing a schedule when work is required to be carried out and by whom, and informing everyone involved.

Levels of provision of site services and facilities

948 While some of the recommended levels of provision in this publication may be reduced for small events, there are areas where a minimum provision will be required. For example, the number of toilets obviously cannot be below two. Realistically, the number of first aiders, stewards, etc, should never be less than two, to allow for contingencies.

Local authority liaison

949 Small-event organisers should consult with the relevant local authority officers and emergency services representatives with responsibility for the event. These officers will be prepared to offer advice and assistance including whether an entertainment licence is required or not.

950 Provide the local authority with sufficient written information to enable officers to understand the nature of the event. This documentation will in any case already have been prepared as part of your event planning.

951 It should include:

- a description of the event, including build-up and breakdown time, audience size, type of activities, etc;
- a site plan showing relevant features and relationship with the neighbourhood;
- a list of key members of the organising team and their responsibilities;
- the risk management strategy, including a copy of the risk assessment, safety policy and site-safety rules.

952 Further documentation should be available on site during the event, including:

- the safety policies, risk assessments and safety method statements for any contractors or subcontractors hired to erect stages, tents, marquees, stalls, etc;
- risk assessments and safety documentation of any activity associated with the entertainment such as bouncy castles, trampolines, etc;
- statutory test certificates for any work equipment brought onto site, such as electrical equipment, generators, lifting equipment.

Chapter 29

Classical music events

953 For the purposes of this chapter, a classical music event is defined as an outdoor performance on a greenfield site - typically parkland within the grounds of a stately home, with an audience who bring with them their own chairs, food and drink and sit where they want within a designated area.

954 As with any event, the initial planning meetings with the local authority and emergency services are critical. It is not uncommon for a touring event to 'build-up' on the morning of a concert and to 'breakdown' immediately on concluding an evening performance.

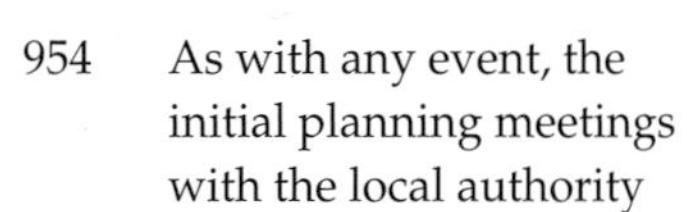

Crowd management

955 Stewarding of a classical concert is very different from a pop concert. The audience profile tends to

be older and less reactive. However, some classical musicians have a following similar to a pop performer and it would be wise to treat their performance as a rock music event.

956 Stewarding may be undertaken (in whole or in part) by local voluntary organisations. These voluntary stewards will require training. An experienced chief steward should be appointed who has been trained to deal with areas of potential conflict (extinguishing lit barbecues, moving audience to avoid over-crowding, etc). A ratio of one steward to 250 audience members has been found to be effective.

Transport management

957 It is unusual for a concert on a greenfield site to be located close to a major transport route. A large proportion of the audience will travel by car and vehicular access via minor estate roads and gated entrances may be a limiting factor on audience capacity. Contingency planning is needed for bad weather conditions; this includes availability of hardcore and tow vehicles and possible re-routing of traffic.

958 It is not uncommon for voluntary stewards to be used to direct traffic on site and organise parking. A more experienced group may be required where the audience exceeds 4000-5000. Voluntary stewards must not direct traffic on or from a public road unless the police have specifically requested it.

Performers

959 A classical orchestra may comprise 75 or more musicians. The addition of choirs can greatly increase this number. It is important that they are provided with dedicated parking and welfare facilities.

Venue and site design

960 When utilising a greenfield site as a music venue, problems arise which would never be encountered in a purpose-built arena. A country estate, originally designed for use by one family, may show signs of strain in coping with a large number of people in one evening. Narrow gateways and steps may become a considerable hazard when used by thousands of people and suitable access routes for site infrastructure (staging, portable toilets, etc) need to be identified. Some hazards, eg lakes, are easily identified. Others, such as rabbit holes, are not, and are only found by tapping local knowledge and walking the site.

961 Livestock may take fright at the sound of music, or the pyrotechnics which almost inevitably, accompany such events. Arrangements should be made to remove livestock before the concert.

962 Evacuation of the site in emergency, and dispersal of the audience, is not normally a problem in open parkland but a venue in a formal garden, with access via gateways, requires careful consideration.

Sanitary facilities

963 The water supply to the site may be a limiting factor on audience size unless re-circulating or non-flush units are employed. A high expectation of the facilities should be anticipated and therefore all units should be serviced throughout the concert.

964 The number of toilets should be based on the recommended standard in the chapter on *Sanitary facilities*. In general, toilets are used more efficiently if they are sited in the same location and easily accessible from the audience area. An exception to this would be wheelchair accessible facilities, which on larger sites could be located on either side of the audience to reduce travel distance.

Food

965 At a typical classical event, the audience will picnic. Personal barbecues are not normally permitted, so the catering facilities will need to serve hot food and drink which the audience cannot provide for themselves.

Waste

966 Greenfield sites are often the home of animals such as deer, sheep and cattle that may suffer considerable harm if the waste is not cleared efficiently. On sites with grazing animals it is important that as much waste as possible is collected on the night of the concert with a sweep the following morning to pick up loose material such as firework debris, nails, bolts and plastic fittings, etc.

967 Additionally, many venues may be open to the public on the day following the concert and it is important that the site is left in the same condition that it was found. Audiences will normally respect the venue and if issued with a rubbish bag (white for visibility after dark) will either take rubbish home or deposit it for collection.

Chapter 30

Unfenced or unticketed events, including radio roadshows

968 Unfenced and/or unticketed events are very popular at open-site venues such as local authority parks. Occasionally free events will be organised in existing arenas or stadia. The aim of this chapter is to highlight specific issues that have to be considered when looking at health and safety at unfenced/unticketed events in open spaces. A few specific suggestions have been made for such events in arenas and stadia and radio roadshows.

Planning and management

969 The planning and management chapter provides information concerning the application of good health and safety management systems.

Risk assessment

970 The whole of the park or open space should be inspected to determine if there are any particular hazards that present greater risks when large numbers of people attend the event. Events that take place next to or with a water feature such as a lake, river or pond will need special provision to prevent people from falling in or swimming in the water. Stewards trained in life-saving skills may need to be employed and extra warning signs erected. In certain circumstances, it may be necessary to physically separate areas of the park or open space from the area chosen for the event.

Build-up/breakdown

971 The fact that there is no perimeter fencing can cause added problems for contractors working on

site. Members of the public will often want to wander around the site to see what is happening. Vehicle movement around the site should keep to dedicated paths and observe strict speed restrictions (5 mph) and use hazard lights. If the park or open space is heavily used it may be necessary to consider having a person walking in front of the moving vehicle.

972 Areas where work is being undertaken can be temporarily cordoned off to ensure that no member of the public wanders into the construction area. Greater security will be needed especially at night to ensure that the temporary structures erected are not vandalised or tampered with in any way. Consideration should therefore be given to the provision of specialist security guards.

973 When erecting temporary demountable structures follow the guidance contained in the chapter on *Structures*. Radio roadshows tend to use vehicles specially adapted for the purpose which contain an integral stage. These vehicles need to be situated on firm level ground that has adequate drainage. If the vehicle is to be placed on grass and there is the possibility of rain, temporary hard standings may need to be considered. It is important to ensure that sufficient space can be provided for the vehicle/stages in the venue design.

Crowd management

974 The benefit of free events held in parks or similar is that there is no enclosed arena so there is no physical restraint to crowd dynamics, however the numbers that are likely to turn up on the day are always difficult to predict. As we have seen throughout the publication, planning for the safety and welfare of the audience is related to the size and nature of the audience attending the event. The numbers of stewards required is dependent upon the overall risk assessment. The fact that the audience members are likely to be spread out over a greater area should be a consideration.

975 In these circumstances, you will need to estimate the expected audience levels. This estimate can vary considerably, dependent upon the popularity of the performers at the time, the weather, other events organised at the same time in the local area and the amount of media attention given to the event. Acquire as much information as possible concerning the expected numbers of audience likely to turn up and all site facilities including stewarding numbers will need to be based on this number. It would be sensible to over-estimate the audience numbers rather than under-estimate.

976 Free or unticketed events organised in existing premises may cause problems when ensuring that the occupant capacity determined for the premises is not exceeded. In these circumstances it may be appropriate to issue free tickets to gain entry to the event or a system for counting audience members in and out of the venue.

A march then the event

977 A march sometimes precedes an event, so that the majority of people arrive at the same time. Care must be taken to ensure that the site and services are ready and able to cope with the large number of people within a limited amount of time. Steward training is essential at this type of event to ensure that the 'audience' is directed to where it is expected.

Crowd information

978 When an event has tickets, information on event times and transportation routes can be given on the reverse of the ticket. When this medium is not available, more emphasis needs to be placed on providing information about the event on leaflets (flyers), local radio, and newspapers. Information could also be made available throughout the event by electronic notice boards.

Major incident planning and emergency access routes

979 Emergency planning and the design of dedicated emergency access routes can prove more difficult at unfenced events as the audience members are not contained in one area. At fenced events, there is relatively easy access around the site once the audience are in the arena watching the event. At unfenced events the audience are able to move to all parts of the park or open space and this can hamper the movement of emergency vehicles. Consider providing cordons with appropriate stewarding to dedicated access routes. Adjustments may need to be made to the existing perimeter fencing of the park to allow for the safe evacuation of the audience from the park, other than through restricted park entrances and exits.

Communication

980 Good communication systems are vitally important to health and safety management. At unfenced events, the location of stewards around the site can be more problematical as there are fewer easily defined stewarding positions, eg entrances and exits to the fenced arena. Stewards need to exhibit greater discipline to remain in the area that they have been stationed and not to wander around the site. Greater reliance on the use of radio communication may be needed at a large site and stewards will need to have clearly gridded plans so that they can be more accurate in summoning assistance and identifying their own position.

Performers

981 It may be necessary to provide a secure backstage area for the performers which is securely fenced to prevent members of the audience trying to get access to the performer. Planning for the arrival and departure of the performer may require cordoning separate areas and roadways.

Children

982 There may be a greater proportion of families with children attending this type of event compared to the traditional ticket/fenced concerts. There is also a higher probability that children and young adults will attend the event on their own. Provide 'help points' and a lost children's facility (see chapter on *Children*).

Information and welfare

983 Provide facilities for information and welfare. The establishment of meeting points for the audience will be more important as there will not be the usual entrances and exit points for audience members to identify with.

Venue and site design

984 Venue design should consider over-spill areas if the numbers that turn up greatly exceed those predicted. Over-spill areas are required to prevent audience members blocking roads or designated emergency escape routes.

985 The numbers of food and merchandising concessions, toilets, first-aid points and other site facilities will depend upon the predicted audience attendance. Careful consideration also needs to be given to the location of the food and merchandising concessionaires, first-aid points, welfare and information points and toilets. It is likely that the audience will be spread over a greater area than is usually calculated for a fenced or enclosed arena. The location of these facilities should reflect this.

Food and drink

986 Glass bottles should not be sold on the site. Local public houses and food outlets should be contacted to request that during the event food and drink is not sold in glass containers.

Waste

987 At unfenced events, it will be effectively impossible to prevent members of the audience taking glass bottles and cans on to the site. Consider providing as much pre-publicity about this aspect as possible. Special containers should be provided to encourage the audience members to dispose of their glass containers safely and if possible encourage people to decant the contents of glass containers into plastic ones.

Chapter 31

All-night music events

988 This chapter highlights some of the matters to consider when organising an all-night music event. All-night events may take place in any of the following: park lands surrounding a stately home, greenfield sites, warehouses, leisure centres/facilities, exhibition halls, purpose-built arenas and night-clubs.

Audience profile

989 All-night events tend to attract an age range of 18-30 years (often with slightly more male attendees). At events more in tune with the concept of 'clubbing', a slightly older audience would be expected: 20-35 years with a 50:50 male:female ratio.

Controlled drugs

990 It may be prudent to arrange for an appropriate drug/alcohol counselling agency to be on site to ensure that if people are in need of advice or assistance, this can be provided. A number of staff should be retained after the event until the site is cleared or no problems are reported.

Admission

991 Consider people who may be queuing in cold weather for long periods of time in the evening or wintertime and make efforts to reduce this time.

Duration of the event

992 All-night events vary in duration, but 10 hours is not unusual for indoor events, with over 16 hours for outdoor weekend events. There will also be a 'build-up' period, followed by 'breakdown' after the show which will impact on the local environment.

Management

993 The arrangements for these events mirror those referred to in other chapters of this publication. Most outdoor events are of a single night's duration but for multiple night events, a night working crew is likely to be necessary to affect 'running repairs' around the site. Due to the duration of these events, ensure that adequate rest is taken by workers and contractors.

Format

994 The format of these events is that different types of music will be played in different areas if the venue layout permits. The audience will move from one area to another throughout the event. This has crowd management implications due to the crowd dynamics of people trying to get into particular locations where the main DJs or live acts are taking place.

995 Ensure that the 'running order' is openly advertised at the venue, eg on the queuing lane, fence panels and at information points. Programming should ensure that crowds are safely distributed around the site according to the capacity of the different areas, to avoid over-crowding, pressure on access points and mass movement around the site.

Venue and site design

996 Outdoor events are susceptible to the weather, and suitable contingency arrangements should be in place, as in severe inclement weather, under-foot conditions can rapidly deteriorate. Low artificial-lighting levels can prevent 'slip and trip' hazards.

997 Vehicle movements around the site should be restricted, particularly during the hours of darkness, as this is potentially very dangerous due to individuals sitting or lying down. Wherever possible, dedicated vehicular routes should be utilised. A 'moat' created between perimeter fencing systems provides an ideal route. However, fence support structures (eg bracing struts), must be clearly highlighted with fluorescent tape or white paint.

Marquees

998 Outdoor events are usually held in dance music tents and/or the open air. An issue of particular concern affecting the audience safety is the availability of tented cover for the occupant capacity of the event. An outdoor daytime concert could last up to 13 hours, with no cover provided against

the weather for the audience. However, at night the air temperature can drop rapidly and with the possibility of limited public transport or remoteness of the site from other facilities, the risk assessment should consider the possibility of hypothermia.

999 A reasonable percentage of the audience should be able to find cover if they so wish, particularly in bad weather. All tented accommodation must comply with the relevant structural and fire safety standards referred to in the chapters on *Structures* and *Fire safety*.

'Chill-out" areas

1000 Fast dancing can result in rising body temperatures in the participants and this can be exacerbated by the effects of some drugs. It therefore is absolutely essential that a 'chill-out' area is provided (or possibly several). This will allow people to cool down, and be in a more calming environment. There may be music, but it will be quieter and more relaxing. It can take a variety of forms, eg a room, tent, marquee with seats, or space outdoors.

1001 If an outdoor 'chill-out' facility is provided in the winter, the air temperature may be so low that heating may be necessary in the area set aside for this use, eg marquee. Stewards should maintain a presence in this area(s), and be particularly observant for individuals who may be in need of medical attention or welfare. If youth/drug counsellors are on site, they should also pay particular attention to these areas. Further information can be found in the publication *Dance till dawn safely*.

Ventilation at indoor venues

1002 Most clubs and warehouse-type venues used for all-night events cannot usually provide an air-conditioned environment. Nonetheless, a variety of methods can be used to achieve a good standard of ventilation.

1003 Due to the vast quantity of hot and humid air that needs to be moved, high velocity fans (forced or induced draught) will be needed to achieve sufficient air changes. If possible, a 'balanced' system should be used where extracted air is replaced by fresh air drawn into the venue. The use of 'smoke' machines or similar effects will need to be carefully assessed.

1004 If the premises cannot be ventilated as previously described, temporary facilities can be brought in to give some relief to the audience, such as high velocity fans placed at approximately 1.6 m (face level). They can be positioned next to areas where individuals will stand, eg bars.

1005 It is preferable if 'cross ventilation' can be provided to move the air quickly. In practice this can be achieved by opening doors or fire exits on opposite sides of the building. Depending on the venue's location the noise 'break-out' may not present a problem; however, in many circumstances it is likely to do so. The opening of any door has to be done on a controlled basis by using the DJs/MCs to explain to the audience the need to briefly reduce sound levels, for safety reasons. If this approach is adopted there should be no public order problems within the venue.

1006 In small arenas or rooms, portable air conditioning units can be used, and will bring about a reasonable reduction in temperature. Lightweight structures will quickly cool in the winter months and will become extremely warm during the summer.

Water

1007 The water supply at events may be provided by a variety of methods. Normally a mains supply will be available but bowsers may have to be used. The water supply inside a permanent venue may be from the rising main or covered/sealed tanks.

1008 It is important to ensure that if a blended water supply is provided at a single tap on the hand-wash basin, the water is not too warm. Hand-wash basin taps which are used for personal hygiene should not be used for drinking water as they can become contaminated in a variety of ways, eg vomit.

1009 Drinking water fountains ensure that waste water can be retained and floors do not become slippery and dangerous. Drinking water taps should always be labelled as such. The pressure of the mains water supply should be adequate for the number of taps (percussion type) being used from it. Seek technical advice form the local water company. One drinking water fountain per 750 people has been found to work in practice.

Free drinking water

1010 This is without doubt the most important aspect of maintaining personal safety at dance events. The individual, particularly at indoor events, may perspire profusely and need an intake of approximately 556 ml (one pint) of fluid per hour. It is essential for the 'core' of the human body to be kept cool, otherwise it will overheat (heat stroke). However a cautionary note: too much water, consumed too quickly can be just as dangerous as too little, and may cause medical problems with serious consequences.

1011 The provision of a free drinking water supply, regardless of the venue type, is an absolute necessity. The method of supplying water can include water bowsers with sparge pipes/taps, drinking water fountains, individual sealed bottles of mineral water and waxed cartons filled by bar or catering staff, etc.

1012 The stewards should know the location of these facilities.

Alcohol and soft drinks

1013 The consumption of alcohol at all-night events varies depending on the nature of the event. High sugar content 'isotonic' soft drinks or fruit juices help to replace body salts and minerals lost through dancing in a hot environment.

Cloakrooms

1014 Throughout the duration of all-night dance events people require different clothing, depending on the temperature. When dancing, people are lightly clothed, and some clothing may be discarded as the event progresses. It is important for a secure cloakroom facility to be available to store bags, jackets, coats, etc.

Welfare

1015 Some additional welfare provisions may be needed. As the majority of people attending such events are young people it is essential to have trained youth or drug workers onsite so that they can identify people who may require support or assistance. These staff should be readily identifiable by the wearing of suitable external identification. They should have an accessible base on site so that they can be easily contacted when their services are needed. A clearly visible meeting point will be required for missing people.

People with special needs

1016 At these events viewing platforms are not usually necessary. Ensure that people with special needs can participate in the event as much as possible. All facilities must be accessible in order that they can maximise their enjoyment. Stewards should be aware of the presence of people with special needs, who may require assistance if an evacuation takes place.

Evacuation

1017 It would be unusual for an evacuation of the whole site to be needed, particularly in a greenfield situation, whereas a single marquee or building may be a more realistic scenario. If possible, a dedicated evacuation area should be used, then after the situation has been resolved and the area is safe, the audience can be allowed back in an orderly manner.

Transport

1018 Special travel arrangements can be put in place, particularly where commercial public transport operators have agreed to provide additional vehicles/rolling stock or change their timetables to suit the event.

1019 For events that are in excess of 12-hours duration, transport arrangements will be necessary to take people back to public transport termini/car parking areas during the event. A proper timetable should be drawn up and as far as possible departure times adhered to. These details should be displayed on information boards or incorporated into handouts available on site. Consideration should be given to finishing the event at a time when public transport has started again in the morning.

Weather forecast

1020 Up-to-date weather information will assist both the production team and the audience. The night temperatures will drop and lightly clothed participants, may find that they suffer medical conditions, eg hypothermia, if they are ill prepared.

Chapter 32

Unlicensed events

1021 The information contained in this publication has primarily been written to provide advice on the application of the Health and Safety at Work etc Act 1974 (HSW Act) at music events and it is usual for such events to require a public entertainment licence from the local authority.

1022 This chapter looks at events that do not require a public entertainment licence. These may include a wide range of entertainment where music is not necessarily the predominant feature. Sporting fixtures, carnivals, fetes and fairs, air shows, agricultural and county shows, events held on public land outside London, and events involving substantial numbers of campers extending over a number of days can fall into this category. There are also specific categories that are excluded by legislation such as religious gatherings and pleasure fairs.

1023 There is no difference between the application of the HSW Act at a licensed or unlicensed event. The difference lies in whether or not the local authority can impose more detailed conditions in relation to public entertainment licensing legislation relevant to the local authority in question. (Entertainment licensing legislation varies between England and Wales and Scotland, in Greater London and other parts of England.)

1024 When organising an event that does not need a public entertainment licence it is not necessary to approach the local authority through the 'licensing' procedures. It is still necessary to comply with the HSW Act and so the information contained in this publication is applicable to unlicensed events.

1025 This publication has been produced for the benefit of event organisers responsible for music events. Elements of this guide can however be used as a template for organising other type of events where large crowds may gather. The *Planning and management* chapter can be applied to most types of public event.

1026 In these circumstances you may find it beneficial to contact the local authority to discuss your proposals and to set up informal discussions with the police, fire authority, ambulance services and other agents that can provide advice on the safe management of the event and advise on compliance with HSW Act and other associated legislation.

1027 It may also be beneficial to agree an informal set of conditions with regard to such aspects as traffic management, signage, security, stewarding, first aid, sanitary facilities, etc, and invite representatives of the local authority, police, fire authority and ambulance services to attend event management meetings. In this way the correct safety information is available to all interested parties.

Chapter 33

Health and safety responsibilities

1028 One of the difficulties often faced by event organisers is determining who has the legal responsibility for the protection of the health, safety and welfare of contractors, self-employed people, suppliers, workers and members of the public on site.

1029 This chapter will outline some of those responsibilities. It should also be noted that this chapter does not attempt to provide an authoritative legal definition of the levels of responsibility for health and safety on site. Legal relationships between promoters, event organisers, contractors, subcontractors, self-employed people and workers can be very complex areas of law. The legal relationships will vary dependent upon the type of contract that is entered into by the respective parties.

Duties of the venue or site owner

1030 Event organisers wishing to stage an event will usually hire a venue or site. The venue could range from a purpose-built stadium or arena owned by a company, individual or local authority to a greenfield site such as a park or collection of fields. These 'owners' may occupy the premises or site themselves or may have granted leases or sub-leases to others. An important factor to consider is who has control over the premises (venue or site). The person in control is defined as the occupier.

1031 A responsible venue or site owner needs to ensure that the venue or site is safe and without risk to anyone who hires the premises and has responsibilities under the following legislation.

Occupiers Liability Act 1957

1032 An occupier of premises owes a 'common duty of care' to all his/her lawful visitors. *The common duty of care is a duty to take such care as in all the circumstances of the case is reasonable to see that the visitor will be reasonably safe using the premises for the purposes for which he/she is invited or permitted by the occupier to be there* (Occupiers Liability Act 1957 section 2 (2)).

Health and Safety at Work etc Act 1974 section 4 (2)

1033 This places a legal duty on people who have control of premises to ensure that the premises, access and exits and any plant and substances within the premises are safe and without risks to the health of people, other than their workers, using the premises as a place of work or a place where they may use plant or substances provided for their use there.

Duties of the event organisers

1034 Music events are usually organised through a promoter. A promoter could be a self-employed person specifically hired by an entertainment agent, artist's manager, record company or other organisation wishing to stage an event. Sometimes different promoters may work together jointly to organise a music event. Promoters may work as event organisers in their own right or employ a production company or contract a self-employed event organiser.

1035 Promoters or venue owners may apply for the entertainment licence in their own name or may request that the event organiser obtains the entertainment licence. The holder of the entertainment licence will be held responsible for breaches in the conditions or requirements of the entertainment licence. Breaches of the Health and Safety at Work etc Act 1974 (HSW Act) and associated regulations may not be so easily defined.

1036 It is therefore crucial that promoters, production companies, event organisers and contractors are clear as to the legal responsibility that each may have in relation to compliance with health and safety legislation. In the majority of circumstances, the responsibility will rest with the event organiser (or stadium management in the case of stadia).

Duties of contractors, subcontractors and self-employed people

1037 A contractor is anyone that has been hired to carry out work who is not an employee. Contractors in turn may hire other subcontractors to carry out part of the work for which they have been contracted.

1038 Contractors and subcontractors as employers have legal duties under the HSW Act of ensuring, so far as is reasonably practicable, the health, safety and welfare of their employees and the health and safety of people not in their employment, who may be affected by their work.

1039 Self-employed people have duties under the HSW Act to ensure that they and anyone else who may be affected by their work, are not exposed to risks to their health.

1040 Contractors, subcontractors, and self-employed people also have duties under the Management of Health and Safety at Work Regulations 1992 (Management Regulations). These include the requirement to have arrangements in place to cover health and safety, assess the risks to workers and other people from their work and to co-operate and exchange information with other employers and self-employed people on site.

1041 Contractors and subcontractors have responsibilities to:

- produce a health and safety policy for their work if five or more people are employed;

- assess the risk to workers and others of their activities. If five or more people are employed, significant findings of the risk assessment must be recorded;
- inform their employees of any risks to their health and safety;
- train their employees;
- provide the correct personal protective equipment for their employees;
- make suitable arrangements for their employees while working on site;
- check the competence of any subcontractors that are hired by them;
- provide information to other employers or self-employed people working on site.

Self-employed people have similar duties and responsibilities that relate to their work.

Employed or self-employed?

1042 The HSW Act section 53 (1), describes a self-employed person as an individual who works for gain or reward otherwise than under a contract of employment, whether or not others are employed by that person. The Inland Revenue definition of a self-employed person depends largely upon the circumstances but it is a person who has a contract for services rather than a contract of service. (A contract of service is the same as a contract of employment.) It is usual to find labour only self-employed subcontractors in the music industry, similar to the construction industry.

1043 Contractors employing labour-only self-employed subcontractors must have clear evidence that these people, even if they pay their own income tax and national insurance contributions, and believe themselves to be self-employed, are in fact self-employed. The absence of a contract of employment, even if these people pay their own income tax and national insurance contributions, is not always sufficient proof.

1044 Many labour-only self-employed people may actually be classified as employees for the manner and type of work that they carry out on site. This aspect is particularly important in relation to the requirement for employers to have Employers Liability Insurance.

Duties of employees

1045 Employees have a duty to take reasonable care for their health and safety and of any other people who may be affected by their acts or omissions at work. They must co-operate with their employer and should not recklessly interfere or misuse anything provided in the interests of health, safety, and welfare. Employees should also notify their employers of any shortcomings in health and safety arrangements.

Standard statement on self-employed people

Although only the courts can give an authoritative interpretation of law, in considering the application of these regulations and guidance to people working under another's direction, the following should be considered:

If people working under the control and direction of others are treated as self-employed for tax and national insurance purposes they may nevertheless be treated as their employees for health and safety purposes. It may therefore be necessary to take appropriate action to protect them. If any doubt exists about who is responsible for the health and safety of a worker this could be clarified and included in the terms of a contract. However, remember, a legal duty under section 3 of the HSW Act cannot be passed on by means of a contract and there will still be duties towards others under section 3 of HSW Act. If such workers are employed on the basis that they are responsible for their own health and safety, legal advice should be sought before doing so.

Legislation

1046 All entertainment events are classed as work activities and are therefore subject to the HSW Act and various regulations and Codes of Practice. In addition, licensing legislation may also apply.

1047 Most health and safety legislation is qualified by the duty to take action 'so far as is reasonably practicable'. 'Reasonably practicable' means that the time, trouble, cost and physical difficulty of taking measures to avoid the risk are not wholly disproportionate to it. The size or financial position of the employer is not taken into account.

1048 Under licensing law, a different concept applies. The licensing authority may impose conditions which are governed by a duty of 'reasonableness'. This means that it can impose requirements which will achieve higher standards than those required under health and safety legislation.

Applying the HSW Act

1049 Event organisers, concert promoters, licensees, specialist contractors and venue owners all have a statutory duty to protect the health and safety of their workers and others who may be affected by their work activity.

Section 2

1050 Section 2 of the Act is concerned with the duties of employers to their employees. The general duty is to ensure, so far as is reasonably practicable, the health, safety and welfare at work of all employees. Some of the most important areas covered by this general duty are specified, eg training, safe systems of work and the preparation of safety policies. This section applies to all organisations with employees at the music event including contractors and construction companies.

Section 3

1051 Section 3 places a duty on employers and self-employed people to safeguard those not in their employment, for example the public. Event organisers should ensure that they are doing all they reasonably can to protect the public. They may have no employees at the venue but their duty to safeguard third parties will extend to providing relevant information to people about aspects of their work which may affect their health and safety, such as emergency procedures. Contractors should consider what effect their work may have on the safety of the employees of other organisations and on the public.

Section 4

1052 Section 4 places a duty on those who have control, to any extent, of non-domestic premises, to ensure as far as is reasonably practicable that the premises are safe and without risks to the health of those who work there. The primary responsibility for the management of risks will usually fall to the event organiser, the manager, the owner of the venue, licensee and/or promoter, depending upon the contractual arrangements under which the event is to be run. The control of the venue may be shared between a number of parties and if this is the case, there should be liaison arrangements to ensure that responsibilities are adequately identified and assigned. Organisations and individuals who have control, to any extent, should consider what measures they can take to ensure that the venue is safe.

Section 6

1053 Section 6 is particularly relevant to suppliers of equipment or substances for use at work. It also applies to designers and suppliers of equipment for use by performers, etc, and to contractors who erect or install stages and sound systems.

Section 7 and 8

1054 Sections 7 and 8 describe broad duties which will apply to all employees at an event.

The Management of Health and Safety at Work Regulations 1992

1055 These Regulations require employers to assess the risks which might exist in the workplace and might affect employees or non-employees (members of the public). They also require them to decide whether safety precautions are adequate and, if they are not, what other control might be needed. Self-employed people must also take similar steps.

1056 Where there are five or more employees, the assessment must be recorded. The record should include the significant findings of the assessment and details of any employees identified as being especially at risk, and what measures are in place to control the risk.

1057 So that employers can carry out their responsibilities properly, the Regulations require that a competent person is appointed to assist them with their health and safety duties.

1058 Where employers share their workplace with another employer or self-employed person, or have another employer's staff working in their premises, they have a duty to co-operate with each other and exchange information on health and safety.

1059 The Regulations also require employers to have procedures in place to deal with serious and imminent danger. This might include evacuation of the workplace. The employer must name a sufficient number of people to put the procedures into practice. They should be trained and competent to carry out their role in an emergency. The evacuation of audience members at an event is usually included in a major incident plan.

The Reporting of Injuries, Diseases and Dangerous Occurrences Regulations 1995

1060 Certain work-related accidents and dangerous occurrences are reportable to the health and safety enforcing authority (either the local authority or HSE depending on the event) under the Reporting of Injuries, Diseases and Dangerous Occurrences Regulations 1995 (RIDDOR).

1061 An employer must report work-related accidents if:

- their employee, or a self-employed person working on premises under their control is killed or suffers certain types of injury;
- a member of the public on premises under their control is killed or taken to hospital; or
- one of the dangerous occurrences listed in the Regulations takes place. These include such incidents as certain scaffold collapses, failure of lifting equipment, certain electrical short circuits.

1062 Further advice about RIDDOR, including the free leaflet *RIDDOR explained* which lists reportable injuries and dangerous occurrences and contains a report form, is available from HSE Books.

Entertainment licensing law

1063 The legislation that may be relevant in relation to licensing law includes:

- Schedule 1 to the Local Government (Miscellaneous Provisions) Act 1982;
- Schedule 12 to the London Government Act 1963;
- Section 41 of the Civic Government (Scotland) Act 1982.

The legislation above have all been amended by Part IV of the Fire Safety and Safety of Places of Sport Act 1987.

1064 Other relevant legislation is as follows:

- The Licensing Act 1964, as amended by the Licensing Act 1988;
- The Licensing (Scotland) Act 1976;
- Private Places of Entertainment (Licensing) Act 1967;
- The Public Entertainment Licence (Drug Misuse) Act 1997; and
- Relevant local acts.

England and Wales

1065 Under the Local Government (Miscellaneous Provisions) Act 1982 or, in Greater London, the London Government Act 1963, responsibility for controlling places which are used for public music and dancing and similar entertainments, including music events, rests with the district council or, in the case of London, the relevant London borough. However, if in England there is no district council, the responsibility rests with the county council. In Wales, responsibility rests only with a county council or a county borough council. It is normally an offence to organise public entertainment without a licence obtained in advance from the local authority, or to be in breach of any of the terms, conditions or restrictions the legislation empowers the authority to place on such a licence.

1066 The purpose of this licensing regime is to ensure, among other things, that places of entertainment have adequate standards of safety and hygiene and to minimise any possible noise which may be caused to the immediate neighbourhood. In considering public entertainment licence applications, the local authority will generally consult with the police and the fire authority, to whom advance notification of the application must be supplied.

1067 The licensing of public entertainment applies to events held indoors throughout the country and to those held outdoors in Greater London. Outside Greater London, licensing applies only to outdoor musical events on private land to which the public have access, and then only if the local authority has taken the necessary steps to adopt the relevant provisions of the 1982 Act.

1068 Local authorities have very wide discretion over whether or not to grant public entertainment licenses and in the case of indoor events or outdoor events held in Greater London, to attach to any licence such terms, conditions and restrictions as they think fit. For outdoor musical events on private land outside Greater London to which the public have access, the local authority may impose terms, conditions and restrictions on any licence it issues only for certain particular purposes which are specified in paragraph 4(4) of Schedule 1 to the Act. These concern:

- securing the safety of performers and other people present at the entertainment;
- ensuring there is adequate access for emergency vehicles and provision of sanitary appliances;
- preventing unreasonable noise disturbance to people in the neighbourhood.

1069 It is, however, possible to impose a variety of terms, conditions or restrictions in respect of an event, provided that they all relate in some way to one of the purposes specified in paragraph 4(4).

1070 In addition, the Private Places of Entertainment (Licensing) Act 1967 enables all local authorities to take on powers, broadly on a similar basis to the wide ranging powers in respect of places of public entertainment, and require the licensing of private events which involve music and dancing that are, for example, promoted for private gain.

Scotland

1071 The council for the local government area may license public entertainments which take place in its area under section 41 of the Civic Government (Scotland) Act 1982, where the particular council has passed a resolution to license a particular class or classes of public entertainment. In such cases, the local authority may grant a public entertainment licence for an event where members of the public attending are required to pay for admission. Where no charge is made, the events cannot be licensed under the 1982 Act. Under section 7 of the 1982 Act, it is an offence to promote an event without a licence in circumstances where a licence is required. In granting such a licence, the local authority may attach conditions regulating such matters as the start and finish time of the event or imposing requirements in respect of tickets, audience capacity, audience density and environmental noise levels. Section 89 of the Act controls the use of temporary seating and staging, etc, and the provision of adequate exits. Such conditions are designed to ensure that there are adequate standards of public safety with respect to the venue, its contents and fire precautions.

1072 An event may also take place on premises for which a licensing board has granted an entertainment licence under the Licensing (Scotland) Act 1976. Such a licence may be granted for places of entertainment such as cinemas, theatres, dance halls and proprietary clubs and permits the sale or supply of liquor for consumption on the premises provided that the sale or supply of liquor is ancillary to a licence to ensure that the sale or supply of alcohol is indeed ancillary to the entertainment.

The Public Entertainment Licence (Drug Misuse) Act 1997

1073 This Act enables a local authority to revoke or not to renew a public entertainment licence if it can be proved that there is a serious problem relating to the supply or use of controlled drugs in connection with the premises. The closure of the premises will take immediate effect and will not be postponed until any appeal by the licencee. The local authority will act on a police report. The local authority may impose terms, conditions or restrictions on the licence such as increasing security measures.

Who enforces?

Health and safety legislation

1074 Enforcement responsibility for health and safety legislation is determined on the basis of 'main activity'. Where the main activity is a leisure activity, which includes a music event, it is the responsibility of the local authority to enforce health and safety legislation, unless the event is organised by the local authority in which case it will be HSE. Certain activities such as radio and television broadcasting and funfairs are retained by HSE. In certain circumstances arrangements are made to transfer the enforcement responsibilities for these activities to the local authority so that they are responsible for the whole event. Health and safety enforcement within local authorities is usually the responsibility of the environmental health officers.

1075 Local authorities and HSE will follow the Health and Safety Commission's policy statement on enforcement. The principles of enforcement include proportionality, consistency, targeting and transparency. Proportionality means relating enforcement action to risks. Consistency means taking a similar approach in similar circumstances to achieve similar ends. Transparency means helping event organisers understand what is expected of them and what they should expect from the enforcing authorities. Targeting means making sure that inspection is targeted primarily on activities that present the most serious risks. Discuss these principles with the health and safety inspector for your event so that you are clear as to what is expected from you and what you can expect from the health and safety inspector.

Entertainment licensing legislation

1076 The enforcement responsibility for entertainment licensing legislation is the responsibility of local authorities who have a statutory responsibility to consult with the police and fire authority. The entertainment licensing function is usually assigned to the Environmental Health Department but may be assigned to different departments within a local authority.

Overlapping legislation

1077 If enforcement action is necessary the decision as to which is the most appropriate legislation will rest with the local authority. Local authorities will therefore ensure that duplicity of enforcement is avoided.

Useful addresses

General

Ambulance Services Association
Friars House
157-168 Blackfriars Road
London SE1 8EU
Tel: 0171 928 9620

Association of Chief Police Officers
Greater Manchester Police
PO Box 22
South West PDO
Chester House
Boyer Street
Manchester M16 0RE
Tel: 0161 872 5050

Association of Festival Organisers
PO Box 296
Aylesbury
Buckinghamshire
HP19 3TL
Tel: 01296 394411

Chief and Assistant Chief Fire Officers Association
10-11 Pebble Close
Amington
Tamworth
Staffordshire B77 4RD
Tel: 01827 61516

Concert Promoters Association
6 St Mark's Road
Henley on Thames
Oxfordshire RG9 1LJ
Tel: 01491 575060

Electrical Contractors Association
ESCA House
34 Palace Court
Bayswater
London W2 4JG
Tel: 0171 313 4800

Entertainment Laser Association
The Old Quarry
Clevedon Road
Failand
Bristol BS8 3TU
Tel: 01275 395050

Football Licensing Authority
27 Harcourt House
19 Cavendish Square
London W1M 9AD
Tel: 0171 491 7191

Institute of Acoustics
77a St Peter's Street
St Albans
Hertfordshire AL1 3BN
Tel: 01727 848195

Institute of Electrical Engineers
Savoy Place
London WC2R 0BL
Tel: 0171 240 1871

Made-Up Textiles Association
42 Heath Street
Tamworth
Staffordshire B79 7JH
Tel: 01827 52337

Mobile Outdoor Caterers Association
Centre Court
1301 Stratford Road
Hall Green
Birmingham B28 9HH
Tel: 0121 693 7000

National Arenas Association
20 Manor Fields
Whalley
Clitheroe
Lancashire BB7 9UD
Tel: 01254 822015

National Outdoor Events Association
7 Hamilton Way
Wallington
Surrey SM6 9NJ
Tel: 0181 669 8121

Production Services Association
Hawks House
School Passage
Kingston -upon-Thames
Surrey KT1 3DU
Tel: 0181 392 0180

Professional Lighting and Sound Association
38 St Leonards Road
Eastbourne
East Sussex BN21 3UT
Tel: 01323 410335

The Chartered Institute of
Building Services
222 Balham High Road
London SW12 9BS
Tel: 0181 675 5211

The Chartered Institute of
Environmental Health
Chadwick Court
15 Hatfields
London SE1 8DJ
Tel: 0171 928 6006

The Event Services Association
8 Home Farm
Ardington
Oxon OX12 8PN
Tel: 01235 821820

The Home Office
50 Queen Anne's Gate
London SW1H 9AT
Tel: 0171 273 3000

The Institution of Structural Engineers
11 Upper Belgrave Street
London SW1X 8BH
Tel: 0171 235 4535

Radio-communications Agency (RA)
New King's Beam House
22 Upper Ground
London SE1 9SA
Tel: 0207 211 0211
Fax: 0207 211 0507

The Royal Environmental Health
Institute of Scotland
3 Manor Place
Edinburgh EH3 7DH
Tel: 0131 225 6999

The Scottish Office
Victoria Quay
Edinburgh EH6 6QQ
Tel: 0131 556 8400

Special needs

Names and addresses of organisations representing people with special needs can be found in *Yellow pages*. For convenience, some of the principal organisations concerned are as follows:

The Access Association
Access Officer
c/o Walsall Metropolitan Borough Council
Civic Centre
Darwall Street
Walsall WS1 1TP
Tel: 01922 652010
Fax: 01922 720885
E-mail: Foxs@walsall.gov.uk

Disabled Living Foundation
380-384 Harrow Road
London W9 2HU
Tel: Helpline 0870 6039177
Fax: 0171 266 2922
E-mail:dlfinfo@dfl.org.uk
Website: www.dlf.org.uk

Joint Committee on Mobility for
Disabled People
c/o Merton and Sutton NHS Trust
Wheelchair Service
7 Damson Way
Fountain Drive
Carshalton
Surrey SM5 4NR
Tel: 0181 770 0693
Fax: 0181 770 0372

Leonard Cheshire Foundation Centre for Accessible Environments
Nutmeg House
60 Gainsford Street
London SE1 2NY
Tel: 0207 357 8182
Fax: 0207 357 8183
Website: www.cae.org.uk
E-mail: cae@globalnet.co.uk

MENCAP
National Centre
123 Golden Lane
London EC1Y 0RT
Tel: 0171 696 5593
Fax: 0171 608 3254,
Website: www.mencap.org.uk
E-mail: mencap.info@dial.pipex.com

MIND
15 -19 Broadway
London E15 4BQ
Tel: Infoline London 0181 522 1728
Outside London 0345 660 163
Website: www.mind.org.uk
E-mail: contact@mind.org.uk

National Federation of the Blind of the United Kingdom
The Old Surgery
215 Kirkgate
Wakefield WF1 1JG
Tel: 01924 291313
Fax: 01924 200244
E-mail: nfbuk@globalnet.co.uk

National Music and Disability Information Service
c/o Sound Sense
Riverside House
Rattlesden
Bury St Edmunds IP30 0SF
Tel: 01449 736287
Fax: 01449 737649
E-mail: 100256.30@compuserve.com

RADAR (the disability network)
Head Office
12 City Forum
250 City Road
London EC1V 8AF
Tel: 0171 250 3222
Fax: 0171 250 0212
E-mail: radar@radar.org.uk

Royal National Institute for the Blind (RNIB)
224 Great Portland Street
London W1N 6AA
Tel: 0171 388 1266
Fax: 0171 388 2034
E-mail: helpline@rnib.org.uk

Royal National Institute for Deaf People (RNID)
19-23 Featherstone Street
London EC1Y 8SL
Tel: 0870 6050123
Fax: 0171 296 8199
Website: http://www.rnid.org.uk

Scottish Accessible Information Forum (SAFE)
Royal Exchange House
100 Queen Street
Glasgow G1 3DN
Tel: 0141 226 5261
Fax: 0141 221 0731

Wales Council for the Disabled
Llys Ifor
Crescent Road
Caerfilly
Mid Glamorgan CF8 1XL
Tel: 01222 887325

Acknowledgements

HSE gratefully acknowledges the assistance of the Home Office and Scottish Office and in particular, the following people in the production of this publication.

Members of the Working Party

Mark Thomas
HSE, Local Authority Unit

Lorraine Miller-Patel
London Borough of Haringey, representing local authorities

Melvin Benn
Mean Fiddler Organisation, representing the Concert Promoters Association (CPA)

Steve Dudley
NEC Group, Birminham, representing the National Arenas Association (NAA)

Richard Limb
Symonds Group Ltd, representing The Events Suppliers Association (TESA) and the National Outdoor Events Association (NOEA)

Tim Norman
ESS, representing the Productions Services Association (PSA)

Dick Tee
Thats EnTeetainment, representing the Productions Services Association (PSA)

Andrew Young
Wembley stadium, representing the National Arenas Association (NAA)

and especially to:

Penny Mellor
Event industry co-ordinator and member of the PSA for help with liaising between HSE and the Event Industry Working Group members, and contributors.

HSE also gratefully acknowledges the participation of the following people in the drafting and updating of this publication. Those people whose names are in bold have written chapters of this publication.

Tim Abbott
Big Green Gathering

Richard Abel
Event Production Management,
The Kayam Theatre Tent

Stephen Abrahall
Festival Information

John Alexander
Scottish Ambulance Service

Steve Anderson
Mendip District Council

Ross Ashton
E/T/C UK Ltd

Andy Ayres
Mantaplan Ltd

Steve Bagnall
London Arena

Richard Balmforth
Austen-Lewis Ltd

Helen Barnsley
Guildford Borough Council

Roger Barrett
Star Hire Ltd

Graham Bawden
Festival Branch

Melvin Benn
Mean Fiddler Organisation

Iain Bisset
Halton Borough Council

Mark Blackstock
Wolverhampton Council

Linda Blair
St Andrew's Ambulance

Mike Boocock
Department of Health

GP Bowles
HM Fire Services Inspectorate

Steve and Janthea Brigden
Nippabout

Peter Brown
Strawberry Fayre

Sally Cavanagh

Lisa Charlwood
Performing Arts Management

Rodney Clark
Pyrovision Ltd

Inspector Alan Clarke
British Transport Police

Tom Clements
Specialized Security

Dick Collins
D C Site Services

John Conway
Reading Borough Council

Paul Cooke
Promed

Jim Cosgrove
London Fire Brigade

Alan Craig
National Arenas Association

Barry Croft
The Royal Borough of Kensington and Chelsea

Dave Crump
Screenco Ltd

Pauline Dalby
Musicians Union

Jenny Davenport
Manchester City Council

Bill Deeker
Pains Fireworks

Jim Dickie

Clive Dickin
Hire Association (Europe) Ltd

Tony Douglas-Beveridge
PLASA

Steve Dudhill
Highway Solutions Ltd

Steve Dudley
NEC Group, Birmingham

Paul Dumpleton
Paul Dumpleton Associates

Roger Duncan

Andrew Dunkley
HSE

Bill Egan
Show Power

Nick Ellison
Stratford on Avon District Council

Kevin Fetterplace
Mojo Working International

Roger Finch
North Hertfordshire Council

Nick Fisher
Firethorn Productions

Keith Flunder
Laser Hire

Bob Fox
MOCA

Andy Frame
TP & TS Ltd

Danni Fuimecelli
London Borough of Islington

Jim Gaffney
Pitstop Barriers

Niall Gaffney
Seating Contracts

Stuart Galbraith
MCP

Geoff Galilee
London Borough of Brent

Aoife Gardiner
The Royal Borough of Kensington and Chelsea

Keith Gosling
London Borough of Brent

John Grant
Walsall Metropolitan Borough Council

John Green
East Lindsey District Council

Jim Griffiths
Symonds Group Ltd

Chris Guy
Playlink

Kevan Habeshaw
Nyrex Arena

Steve Haddrell
WOMAD

David Hall
London Borough of Tower Hamlets

Chris Hannam
South Western Management

Les Hart
Home Office

Steve Heap
Assocation of Festival Organisers

Barbara Herridge
UK Waste

Mike Herriot
Scottish Ambulance Service

Steve Hick
Home Office Emergency Planning College

Peter Hind
Total Fabrications

Dr Ken Hines
BASICS

Judy Hoatson
Reading Borough Council

MS Hooker
Solihull Metropolitan Borough Council

Dr Chris Howes
Festival Medical Services

Roy Hunt
Hunts Exhibition Services

Penny Jackson
National Youth Arts Festival

Alan Jacobi
Unusual Services

Keith James
Cardiff City Council

Fiona Jones
Cambridge City Council

Frank Jones
Hertfordshire Fire and Rescue

Paul Kilgallen
Leicestershire Ambulance Service

Stewart Kingsley
Chelmsford Borough Council

Liz Kwast

Dr C Laird
BASICS (Scotland)

Ray and Bev Langton

Gary Lathan
Glasgow City Council

Peter and Katrina Lawrence
Celebrate Independant Promoters Association

Richard Limb
Symonds Group Ltd

Charles Lister
Search Ltd

Bethany and Tony Llewellyn Creek
Fairs and Festivals Federation

Bernard Lloyd
Central Catering

Sergeant Gary Lockyer
Metropolitan Police

Paul Ludford

Hash Maitra
HSE

Mike Mathieson
Cake

Penny Mellor

Commander Mike Messinger
Metropolitan Police

David Miller
Gofer

Lorraine Miller-Patel
Candlish Miller Consultants

Kevin Minton
Hire Association (Europe) Ltd

Steve Newman
Stratford on Avon District Council

Tim Norman
ESS

Peter O'Conner
Central Office of Information

John O'Hagan
NRPB

Shawn O'Malley
WAVE

Rachel Parker
Earls Court

Lincoln Parkhouse
Just FX

Graham Pollock
West Dumbarton Council

Dave Pratley
Helter Skelter

Ray Rhodes
RTW Concessions Ltd

Walter Richardson
British Red Cross

Andy Rock
Network Recycling

Peter Rooke
Cherwell District Council

Tony Rosenburg
Sanctuary Leisure

Richard Saunders
Royal London Borough of Kensington and Chelsea

Paul Scott
Belfast City Council

Wilf Scott
Pyrovision Ltd

Robert Seaman
Oxfordshire Ambulance Service

Donald Sinclair

Michael Skelding
MUTA

Peter Smith
Crawley Borough Council

Mark Stracey
Bravado/CMI

Bill Stuart
Stuart Leisure and Security

Tony Sullivan
London Fire and Civil Defence Authority

Peter Swindlehurst
St John's Ambulance

David Taylor
London Ambulance Service

Heather Taylor
Chelmsford Borough Council

Dick Tee
Thats EnTeetainment

David Tolley
London Borough of Tower Hamlets

Mary Treacy

Steve Tuck
Blackout

Mick Upton
Showsec

Brian Waddingham
London Borough of Haringey

Tony Wadley
TESS

Steve Walker
HSE

Annie Watson

G Weaver
London Borough of Islington

Alan Webb
HSE

Gary White
Unusual Engineering Ltd

Dick Whittingham
Wembley Arena

Colin Wickes
London Borough of Brent

Ray Williams
Williams Management Communications

Geoff Wilson
Football Licensing Authority

Philip Winsor
Milton Keynes Borough Council

Dave Withey
Arena Seating

Nich Woolf
Festival Medical Services

Andrew Young
Wembley Arena

References

The references have been arranged so that the titles are arranged alphabetically.

A guide to audiometric testing programmes MS26 HSE Books 1995 ISBN 0 7176 0942 1

A guide to the Reporting of Injuries, Diseases and Dangerous Occurrences Regulations 1995 L73 HSE Books 1996 ISBN 0 7176 1012 8

Avoidance of danger from overhead electricity lines GS6(rev) HSE Books 1997 ISBN 0 7176 1348 8

Avoidance of danger from underground services HSG47 HSE Books 1998 ISBN 0 7176 0435 7

Carriage of Dangerous Goods (Amendment) Regulations 1999 SI 1999/303 ISBN 0 11 080470 8

Carriage of Dangerous Goods by Road (Driver Training) Regulations 1996 (DTR) HMSO 1996 ISBN 0 11 0629280

Carriage of Explosives by Road Regulations 1996 SI 1996/2093 HMSO 1996 ISBN 0 11 062925 6

Civic Government (Scotland) Act 1982 Ch 45 1982 HMSO ISBN 0 10 544582 7

Classification and Labelling of Explosives Regulations 1983 SI 1983/1140 HMSO 1983 ISBN 0 11 037140 2

Code of Practice on Environmental Noise Control at Concerts Noise Council 1995 ISBN 0 900103 51 5

Conditions for the authorisation of explosives in Great Britain HSG114 HSE Books 1994 ISBN 0 7176 0717 8

Control of Explosives Regulations 1991 SI 1991/1531 HMSO 1991 ISBN 0 11 014531 3

Controlled Waste Regulations 1999 SI 1992/588 HMSO 1992 ISBN 0 11 023588 6

Controlling the radiation safety of display laser installation INDG224 HSE Books 1996

London Drug Policy Forum *Dance till dawn safely: A Code of Practice on health and safety at dance venues* LDPF 1996

Home Office *Dealing with disaster* 3rd ed Brodie Publishing 1997 ISBN 185 983 9208

Disability Discrimination Act 1995 Ch 50 HMSO 1995 ISBN 0 10 545095 2

HELA *Disco lights and flicker sensitive epilepsy* LAC51/1 HSE 1996

Electrical safety at places of entertainment GS50 2nd ed HSE Books 1997 ISBN 0 7176 1387 9

Electrical safety for entertainers INDG247 HSE Books 1997

Environment Act 1995 Ch 25 HMSO 1995 ISBN 0 10 542595 8

Environmental Protection Act 1990 Ch 43 *Duty of care: A Code of Practice* HMSO 1990 ISBN 0 10 544390

Everyone's guide to RIDDOR95: Reporting of Injuries, Diseases and Dangerous Occurrences Regulations 1995 HSE Books 1996 ISBN 0 7176 1077 2

Explosives Act 1875 Ch 17 HMSO 1875

Fairgrounds and amusement parks: Guidance on safe practice. Practical guidance on the management of health and safety for those involved in the fairgrounds industry HSG175 HSE Books 1997 ISBN 0 7176 1174 4

Fire Precautions (Workplace) Regulations 1997 SI 1997/1840 HMSO 1997 ISBN 0 11 064738 6

Fire Precautions Act 1971 Ch 40 HMSO 1971 ISBN 0 10 544071X

Fire Safety and Safety of Places of Sport Act 1987 Ch 27 HMSO 1987 ISBN 0 10 542787X

Fireworks (Safety) Regulations 1997 SI 1997/2294 HMSO 1997 ISBN 0 11 064962 1

First aid at work: The Health and Safety (First-Aid) Regulations 1981 Approved Code of Practice and Guidance L74 HSE Books 1997 ISBN 0 7176 1050 0

Five steps to risk assessment: A step by step guide to a safer and healthier workplace INDG163 HSE Books 1998 ISBN 0 7176 0904 9

General COSHH ACoP (Control for Substances Hazardous to Health) and Carcinogens ACoP (Control of Carcinogenic Substances) and Biological Agents ACoP (Control of Biological Agents) Control of Substances Hazardous to Health Regulations 1999. Approved Codes of Practice HSE Books 1999 ISBN 0 7176 1670 3

Guide to fire precautions in existing places of entertainment and like premises Home Office and Scottish Home and Health Department 1990 ISBN 0 11 340907 9

Guide to safety at sports grounds 4th ed Department of National Heritage and Scottish Office 1997 ISBN 0 11 300095 2

Health and Safety (First Aid) Regulations 1981 SI 1981/917 HMSO 1981 ISBN 0 11 016917 4

Health and Safety at Work etc Act 1974 CH 37 HMSO 1974 ISBN 0 10 543774 3

Health surveillance in noisy industries: Advice for employers INDG193 HSE Books 1995 ISBN 0 7176 0933 2

Licensing (Scotland) Act 1976 Ch 66 HMSO 1976 ISBN 0 10 546676X

Licensing Act 1964 Ch 26 HMSO 1964 ISBN 0 10 850263 5

Licensing Act 1988 Ch 17 HMSO 1988 ISBN 0 10 541788 2

Local Government (Miscellaneous Provisions) Act 1982 Ch 30 HMSO1982 ISBN 0 10 543082X

London Government Act 1963 Ch 33 HMSO 1963 ISBN 0 10 850313 5

Maintaining portable and transportable electrical equipment HSG107 HSE Books 1994 ISBN 0 7176 0715 1

Management of health and safety at work Management of Health and Safety at Work Regulations 1992: Approved Code of Practice L21 HSE Books 1992 ISBN 0 7176 0412 8

Managing construction for health and safety Construction (Design and Management) Regulations 1994 Approved Code of Practice L54 HSE Books 1995 ISBN 0 7176 0792 5

Managing contractors: A guide for employers An open learning booklet HSE Books 1997 ISBN 0 7176 1196 5

Managing crowds safely HSG154 HSE Books 1996 ISBN 0 7176 1180 9

Manual handling Manual Handling Operations Regulations 1992 Guidance on regulations L23 HSE Books 1992 ISBN 0 7176 0411 X

Memorandum of guidance on the Electricity at Work Regulations 1989 HSR25 HSE Books 1989 ISBN 0 7176 1602 9

Occupational exposure limits 1998 EH 40 HSE Books 1998 ISBN 0 7176 1474 3

Occupiers Liability Act 1957 Ch 31 HMSO 1957 ISBN 0 10 50198 1

Privy Council *Order in council relating to relating to stores licenced for mixed explosives* Order in council no 6 HMSO 1875

Privy Council *Order in council amending order in council (no.6) of the 27th day of November 1875, relating to stores licensed for mixed explosives* Order in council no 6a HMSO 1883

Privy Council *Order in council relating to premises registered for the keeping of mixed explosives* Order in council no 16 HMSO 1896

Packaging of Explosives for Carriage Regulations 1991 SI 1991/2097 HMSO 1991 ISBN 0 11 015097 X

Personal protective equipment at work Personal Protective Equipment at Work Regulations 1992 Guidance on regulations L25 HSE Books 1992 ISBN 0 7176 0415 2

Placing on the Market and Supervision of Transfers of Explosives Regulations 1993 (POMSTER) SI 1993/2714 HMSO 1993 ISBN 0 11 035714 0

Private Places of Entertainment (Licensing) Act 1967 Ch 19 HMSO 1967 ISBN 0 10 850437 9

Public Entertainment Licences (Drug Misuse) Act 1997 Ch 49 HMSO 1997 ISBN 0 10 544997 0

Radiation safety of lasers used for display purposes HSG95 HSE Books 1996 ISBN 0 7176 0691 0

Reducing noise at work: Guidance on the Noise at Work Regulations 1989 HSE Books 1989 ISBN 0 7176 1511 1

Research to develop a methodology for the assessment of risks to crowd safety in public venues Parts 1 and 2 Contract research report no 204/1998 HSE Books 1999 ISBN 0 7176 1663 0

RIDDOR explained: A short guide to the Reporting of Injuries, Diseases and Dangerous Occurrences Regulations 1995 HSE31(rev1) HSE Books 1999

Rider-operated lift trucks: Operator training Approved Code of Practice and guidance L117 HSE Books 1999 ISBN 0 7176 2455 2

Safe operation of passenger carrying amusement devices Inflatable bouncing devices PM76 HSE Books 1991 ISBN 0 11 885604 9

Safe operation of passenger carrying amusement devices The ark/speedways PM 70 HSE Books 1988 ISBN 0 11 885407 0

Safe operation of passenger carrying amusement devices The big wheel PM57 HSE Books 1986 ISBN 0 11 883536 X

Safe operation of passenger carrying amusement devices The chair-o-plane PM61 HSE Books 1986 ISBN 0 11 883928 4

Safe operation of passenger carrying amusement devices The cyclone twist PM49 HSE Books 1985 ISBN 0 11 883525 4 Out of print

Safe operation of passenger carrying amusement devices The octopus PM48 HSE Books 1985 ISBN 0 11 883607 2

Safe operation of passenger carrying amusement devices The paratrooper PM59 HSE Books 1986 ISBN 0 11 883534 3

Safe operation of passenger carrying amusement devices The roller coaster PM68 HSE Books 1987 ISBN 0 11 883942 X

Safe operation of passenger carrying amusement devices The trabant PM72 HSE Books 1990 ISBN 0 11 885424 0

Safe operation of passenger carrying amusement devices The waltzer PM47 HSE Books 1985 ISBN 0 11 883608 0 Out of print

Safe operation of passenger carrying amusement devices The water chute PM71 HSE Books 1989 ISBN 0 11 885415 1

Safe use of lifting equipment Lifting Operations and Lifting Equipment Regulations 1998 (LOLER) Approved Code of Practice and guidance L113 HSE Books 1998 ISBN 0 7176 1628 2

Safe use of work equipment Provision and Use of Work Equipment Regulations 1998 (PUWER) Approved Code of Practice and guidance L22 2nd ed HSE Books 1998 ISBN 0 7176 1626 6

Safety of Sports Grounds Act 1975 Ch 52 HMSO 1975 ISBN 0 10 545275 0

Safety signs and signals Health and Safety (Safety Signs and Signals) Regulations 1996 Guidance on regulations L64 HSE Books 1996 ISBN 0 7176 0870 0

Smoke and vapour effects used in entertainment ETIS 3 HSE Books 1996

Special Waste Regulations 1996 SI 1996/972 HMSO 1996 ISBN 0 11 054565 6

Barnett P and Woodgate J *Stadium public address systems* Football Stadia Advisory Design Council 1991 ISBN 1 87 383110 2

Temporary demountable structures: Guidance on design, procurement and use 2nd ed Institution of Structural Engineers 1999 ISBN 1 874266 45 X

The BERSA Code of Safe Practice British Elastic Rope Sports Association 1993

The Children Act 1989 Ch 41 HMSO 1989 ISBN 0 10 544189 9

Writing plain English: A guide for writers and designers of official forms, leaflets, letters, labels and agreements Plain English Campaign 1980 ISBN 0 907 42400 7

Waste Management Licensing Regulations 1994 SI 1994/1056 HMSO 1994 ISBN 0 11 044056 0

Working at heights in the broadcasting and entertainment industries ETIS6 HSE Books 1998

Working together on firework displays A guide to safety for display organisers and operators HSG123 HSE Books 1995 ISBN 0 7176 0835 2

Workplace health, safety and welfare Workplace (Health, Safety and Welfare Regulations) 1992 Approved Code of Practice and guidance L24 HSE Books 1992 ISBN 0 7176 0413 6

Writing a safety policy statement: Advice for employers HSC6 HSE Books 1985

While every effort has been made to ensure the accuracy of the references listed in this publication, their future availability cannot be guaranteed.

British Standards

The British Standards are arranged numerically.

BS 3169: 1986 *Specification for first-aid reel hoses for fire fighting purposes*

Colour and diffusion filter material for theatre, television and entertainment purposes
BS 3944: Part 1 1992 *Part 1 Specification for flammability and dimensional stability*

Emergency lighting
BS 5266: Part 1 1988 *Part 1 Code of Practice for emergency lighting of premises other than cinemas and other specified premises for entertainment*
BS 5266: Part 2 1988 *Part 2 Code of Practice for electrical low mounted way guidance systems for emergency use*
BS 5266: Part 3 1988 *Part 3 Emergency lighting for small, powered relays (electromagnetic) for emergency lighting applications up to and including 32 A*

BS 5274 *Specification for fire hose reels (water) for fixed installation purposes* (withdrawn replaced by) *Fixed fire fighting systems Hose systems*
BS EN 671-1: Part 1 1995 *Part 1 Hose reels with semi-rigid hose*

Fire extinguishing installations and equipment on premises
BS 5306: Part 1 1976 *Part 1: Fire extinguishing installations and equipment on premises: hydrant systems, hose reels and foam inlets*
BS 5306: Part 3 1985 *Part 3: Code of practice for selection, installation and maintenance of portable fire extinguishers*

BS 5438: 1989 (amd 1995) *Methods of test for flammability of textile fabrics when subjected to a small igniting flame applied to the face or bottom edge of vertically oriented specimens*

Fire safety signs, notices and graphic symbols
BS 5499 Part 1: 1990 (amd 1995) *Part 1: Specification for fire safety designs*
BS 5499 Part 2: 1986 (amd 1995) *Part 2: Specification for self-illuminating fire safety signs*
BS 5499 Part 3: 1990 *Part 3: Specification for internally illuminated fire safety signs*

BS 5696 Parts 1-3 *Play equipment intended for permanent installation outdoors* (withdrawn replaced by) *Play equipment*
BS EN 1176: Part 1 1998 *Part 1 General safety requirements and test methods*
BS EN 1176: Part 2 1998 *Part 2 Additional specific safety requirements and test methods for swings*
BS EN 1176: Part 3 1998 *Part 3 Additional specific safety requirements and test methods for slides*
BS EN 1176: Part 4 1998 *Part 4 Additional specific safety requirements and test methods for runways*
BS EN 1176: Part 6 1998 *Part 6 Additional specific safety requirements and test methods for rocking equipment*
BS EN 1176: Part 7 1997 *Part 7 Guidance on installation, inspection, maintenance and operation*

BS 5810: 1979 *Code of practice for access for the disabled to buildings*

Fire detection and alarm systems for buildings
BS 5839: Part 1 1998 *Part 1: Code of practice for system design, installation and servicing*
BS 5839: Part 2 1983 *Part 2: Specification for manual call points*
BS 5839: Part 3 1988 *Part 3: Specification for automatic release mechanisms for certain fire protection equipment*
BS 5839: Part 4 1988 *Part 4: Specification for control and indicating equipment*
BS 5839: Part 5 1988 *Part 5: Specification for optical beam smoke detectors*
BS 5839: Part 6 1995 *Part 6: Code of practice for the design and installation of fire detection and alarm systems*
BS 5839: Part 8 1998 *Part 8: Code of practice for the design, installation and servicing of voice alarm system*

Specification for fabrics for curtains and drapes
BS 5867: Part 1 1980 (amd 1993) *Part 1: General requirements*
BS 5867: Part 2 1980 (amd 1993) *Part 2: Flammability requirements*

BS 6465: Part 1 1994 *Part 1: Code of practice for scale of provision, selection and installation of sanitary appliances*

BS 6472: 1992 *Guide to the evaluation of human exposure to vibration in buildings* (1 Hz-80 Hz)

BS 6575: 1985 *Specification of fire blankets*

Fireworks
BS 7114 Part 1: 1988 *Part 1: Classification of fireworks*
BS 7114 Part 2: 1988 *Part 2: Specification for fireworks*
BS 7114 Part 3: 1988 *Part 3: Methods of test for fireworks*

Evaluation and measurement for vibrations in buildings
BS 7385: Part 1 1990 *Part 1: Guide for measurement of vibration and evaluation of their effects on buildings*
BS 7385: Part 2 1993 *Part 2: Guide to damage from groundborne vibration*

BS 7430: 1991 *Code of Practice for earthing*
BS 7671: 1992 *Requirements for electrical installations*

BS 7863: 1996 *Recommendations for colour coding to indicate the extinguishing media contained in portable fire extinguisher*

BS 7909: 1998 *Code of Practice for temporary distribution systems for ac electrical supplies for entertainment, lighting, technical services and related purposes*

BS EN 2: 1992 *Classification of fires*

Portable fire extinguishers
BS EN 3: Part 1 1996 *Part 1: Description, duration of operation, class A and B fire test*
BS EN 3 : Part 2 1996 *Part 2: Tightness, dielectric test, tamping test, special provisions*

BS EN 179: 1998 *Building hardware Emergency exit devices operated by lever handle or push pad Requirements and test methods*

BS EN 1125: 1997 *Building hardware Panic exit devices operated by a horizontal bar Requirements and test methods*

Safety of laser products
BS EN 60825: Part 1 1994 *Part 1: Equipment classification, requirements and user's guide*

BS EN 60849: 1998 *Sound systems for emergency purposes*

Further reading

The *Further reading* references have been arranged in alphabetical order according to their titles.

A short guide to the Personal Protective Equipment at Work Regulations 1992 INDG174 HSE Books 1997 ISBN 0 7176 0889 1

A step-by-step guide to COSHH assessment HSG97 HSE Books 1993 ISBN 0 7176 1446 8

Assured safe catering: A management system for hazard analysis 1993 Department of Health ISBN 0 11 321688 2

Camera operations on location: Guidance for managers and camera crews HSG169 HSE Books 1997 ISBN 0 7176 1346 1

Construction (Head Protection) Regulations 1989 Guidance on Regulations L25 HSE Books 1992 ISBN 0 11 885503 4

COSHH and peripatetic workers HSG77 HSE Books 1992 ISBN 0 11 885733 9

Electrical safety and you INDG231 HSE Books 1996 ISBN 0 7176 1207 4

Electricity at work: Safe working practices HSG85 1993 ISBN 0 7176 0442 X

Fire safety: An employer's guide HSE Books 1999 ISBN 0 11 34122 9

Scottish Drugs Forum *Guidelines for good practice at dance events* 1995 Scottish Drugs Forum ISBN 0 9519761 2 5

An index of health and safety guidance for the catering industry CAIS7 HSE Books 1996

Association of Chief Police Officers General Policing Committee Standing Sub-committee on Emergency Planning *Emergency procedures manual* 1997 ACOP

Industry guide to good food hygiene practice: Catering guide Chadwick House Group 1997 ISBN 0900 103 00 0

Industry guide to good food hygiene practice: Markets and fairs guide Chadwick House Group 1998 ISBN 1 902 42300 3

HELA *Keeping of LPG in vehicles: Mobile catering units* 52/13 HSE Books 1986

LP Gas Association *Use of LPG cylinders in mobile catering vehicles and similar commercial units* Code of Practice 24 Part 3 LP Gas Association 1996 ISBN 1 87 39118 0

Au SYZ and Ryan MC *Managing crowd safety in public places: A study to generate guidance for venue owners and enforcing authority inspectors* CRR53 HSE Books 1993 ISBN 0 7176 0780 9

Managing health and safety: Five steps to success INDG275 HSE Books 1998

Managing vehicle safety at the workplace: Leaflet for employers INDG199 HSE Books 1995 ISBN 0 7176 0982 0

MOCA Code of practice for mobile and outside caterers 2nd ed The Mobile Outdoor Caterer Association 1999

MOCA Due diligence system 2nd ed The Mobile Outdoor Caterer Association 1999

National Outdoor Events Association *Code of Practice for outdoor events* 1993 NOEH plus amendments 1997

Protecting the public: Your next move HSG151 HSE Books 1997 ISBN 0 7176 1148 5

Reversing vehicles INDG148 HSE Books 1993 ISBN 0 7176 1063 2

Safe erection of structures Part 2 Site management and procedures GS28/2 HSE Books 1985 ISBN 0 11 883605 6

Safe erection of structures Part 3 Working places and access GS38/3 HSE Books 1986 ISBN 0 11 883530 0

Safety in working with lift trucks HSG6 HSE Books 1992 ISBN 0 11 886395 9

Selecting a health and safety consultancy INDG133 HSE Books 1992

Football League, Football Association and Football Association Premier League *Stewarding and safety management at football grounds* 1995

Successful health and safety management HSG65 HSE Books 1991 ISBN 0 7176 0425 X

Technical standards for marquees and large tents DOC14 Home Office 1995

The keeping of LPG in cylinders and similar containers CS4 HSE Books 1986 ISBN 0 7176 0613 7

Workplace transport safety: Guidance for employers HSG136 HSE Books 1995 ISBN 0 7176 0935

Index

Note: page numbers for major topics are **emboldened**.

Printed and published by the Health and Safety Executive
C30 4/01